Daphne Isabel Jost

Geheimkot

Daphne Isabel Jost

Geheimkot

Fütterungsbedingte mikrobielle Zusammensetzung von Rinderkot, Einfluss auf Gasemissionen und N-Mineralisierung im Boden

Südwestdeutscher Verlag für Hochschulschriften

Impressum/Imprint (nur für Deutschland/only for Germany)
Bibliografische Information der Deutschen Nationalbibliothek: Die Deutsche Nationalbibliothek verzeichnet diese Publikation in der Deutschen Nationalbibliografie; detaillierte bibliografische Daten sind im Internet über http://dnb.d-nb.de abrufbar.
Alle in diesem Buch genannten Marken und Produktnamen unterliegen warenzeichen-, marken- oder patentrechtlichem Schutz bzw. sind Warenzeichen oder eingetragene Warenzeichen der jeweiligen Inhaber. Die Wiedergabe von Marken, Produktnamen, Gebrauchsnamen, Handelsnamen, Warenbezeichnungen u.s.w. in diesem Werk berechtigt auch ohne besondere Kennzeichnung nicht zu der Annahme, dass solche Namen im Sinne der Warenzeichen- und Markenschutzgesetzgebung als frei zu betrachten wären und daher von jedermann benutzt werden dürften.

Coverbild: www.ingimage.com

Verlag: Südwestdeutscher Verlag für Hochschulschriften GmbH & Co. KG
Heinrich-Böcking-Str. 6-8, 66121 Saarbrücken, Deutschland
Telefon +49 681 37 20 271-1, Telefax +49 681 37 20 271-0
Email: info@svh-verlag.de

Zugl.: Witzenhausen, Universität Kassel, Diss., 2011

Herstellung in Deutschland:
Schaltungsdienst Lange o.H.G., Berlin
Books on Demand GmbH, Norderstedt
Reha GmbH, Saarbrücken
Amazon Distribution GmbH, Leipzig
ISBN: 978-3-8381-3158-0

Imprint (only for USA, GB)
Bibliographic information published by the Deutsche Nationalbibliothek: The Deutsche Nationalbibliothek lists this publication in the Deutsche Nationalbibliografie; detailed bibliographic data are available in the Internet at http://dnb.d-nb.de.
Any brand names and product names mentioned in this book are subject to trademark, brand or patent protection and are trademarks or registered trademarks of their respective holders. The use of brand names, product names, common names, trade names, product descriptions etc. even without a particular marking in this works is in no way to be construed to mean that such names may be regarded as unrestricted in respect of trademark and brand protection legislation and could thus be used by anyone.

Cover image: www.ingimage.com

Publisher: Südwestdeutscher Verlag für Hochschulschriften GmbH & Co. KG
Heinrich-Böcking-Str. 6-8, 66121 Saarbrücken, Germany
Phone +49 681 37 20 271-1, Fax +49 681 37 20 271-0
Email: info@svh-verlag.de

Printed in the U.S.A.
Printed in the U.K. by (see last page)
ISBN: 978-3-8381-3158-0

Copyright © 2012 by the author and Südwestdeutscher Verlag für Hochschulschriften GmbH & Co. KG and licensors
All rights reserved. Saarbrücken 2012

Mis queridos

Die vorliegende Arbeit wurde vom Fachbereich Ökologische Agrarwissenschaften der Universität Kassel als Dissertation zur Erlangung des akademischen Grades eines Doktors der Naturwissenschaften (Dr. rer. nat.) angenommen.

Erstgutachter: Prof. Dr. A. Sundrum
Zweitgutachter: Prof. Dr. R.G. Jörgensen

Tag der mündlichen Prüfung: 2. Dezember 2011

Vorwort

Die vorliegende Dissertation wurde im Rahmen des DFG-Graduiertenkollegs 1397 an der Universität Kassel im Fachbereich Ökologische Agrarwissenschaften im Fachgebiet Bodenbiologie und Pflanzenernährung angefertigt, um die Anforderungen des akademischen Grades des Doktors der Naturwissenschaften (Dr. rer. nat.) zu erfüllen. Die Arbeit besteht aus drei wissenschaftlichen Publikationen. Die erste wurde bereits bei einer international begutachteten Fachzeitschrift veröffentlicht, die zweite und dritte stehen kurz vor der Einreichung. Die Artikel sind in Kapitel 4, 5 und 6 eingearbeitet. Kapitel 1 liefert eine generelle Einleitung zum Thema, während in Kapitel 2 die Ziele dieser Arbeit herausgestellt werden. Die methodische Herangehensweise, Vorversuche und nicht verwendete Ergebnisse werden im 3. Kapitel dargestellt. Kapitel 7 fasst die Ergebnisse der Kapitel 4, 5 und 6 zusammen, und wird in Kapitel 8 in Englisch aufgeführt. Eine Schlussfolgerung, die sich aus den Untersuchungen dieser Arbeit ergibt, bildet Kapitel 9. Ein Ausblick für weitere Untersuchungen findet sich in Kapitel 10. In Kapitel 11 finden sich die Quellen, die für die Kapitel 1, 2, 3 und 9 benötigt wurden.

Folgende Publikationen sind in die Arbeit eingebettet:

Kapitel 4

Jost, D. I., Indorf, I., Joergensen, R.G., Sundrum, A. (2011). The determination of microbial biomass in cattle faeces using soil microbiological methods. Soil Biology and Biochemistry 43, 1237-1244.

Kapitel 5

Jost, D. I., Aschemann, M., Lebzien, P., Joergensen, R.G., Sundrum, A. Microbial biomass in faeces of dairy cows affected by a nitrogen deficient diet. (to be submitted to Animal Feed Science and Technology).

Kapitel 6

Jost, D. I., Joergensen, R.G., Sundrum, A. Effect of cattle faeces with different microbial biomass content on soil properties, gaseous emissions and plant growth. (to be submitted to Journal of Agricultural Science)

Inhaltsverzeichnis

Abbildungsverzeichnis
Tabellenverzeichnis
Abkürzungsverzeichnis

1. Einleitung 1

   1.1 Steuerung des Nährstoffhaushalts in der Ökologischen Landwirtschaft 1

   1.2 Biologische Indikatoren der Bodenfruchtbarkeit 2

   1.3 N_2O-Emissionen aus der Landwirtschaft 3

2. Ziele der Arbeit 4

3. Methodik und Vorversuche 5

   3.1 Chloroform-Fumigations-Extraktionsmethode (CFE) 5

   3.2 Ergosterolbestimmung 5

   3.3 Aminozuckerbestimmung 6

   3.4 Luminometrische ATP-Bestimmung 7

   3.5 Probenkonservierung 8

4. Determination of microbial biomass and fungal and bacterial distribution in cattle faeces 10

   4.1 Introduction 11

   4.2 Materials and methods 13

      4.2.1. Faeces sampling, quality determination, and incubation 13

      4.2.2. Microbial biomass C and N estimation by CFE 14

      4.2.3. Ergosterol analysis 15

      4.2.4. Amino sugar analysis 15

      4.2.5. Statistical analysis 16

- 4.3 Results 16
  - 4.3.1. Extractant efficiency, variability, and stability of the CFE method 16
  - 4.3.2. Microbial biomass and ergosterol in incubated faeces 18
  - 4.3.3. Relationships of microbial indices in faeces of different heifers 19
- 4.4 Discussion 20
  - 4.4.1. Microbial biomass in cattle faeces by the CFE method 20
  - 4.4.2. Bacterial and fungal distribution in cattle faeces 21
  - 4.4.3. Conclusions 24
- 4.5 References 24

5. Microbial biomass in faeces of dairy cows affected by a nitrogen deficient diet 31
  - 5.1 Introduction 32
  - 5.2 Materials and Methods 33
    - 5.2.1. Feeding regime 33
    - 5.2.2. Faecal C and N fractions 34
    - 5.2.3. Microbial biomass C and N 35
    - 5.2.4. Ergosterol analysis 35
    - 5.2.5. Amino sugars 36
    - 5.2.6. Statistical analysis 36
  - 5.3 Results 37
  - 5.4 Discussion 41
    - 5.4.1. Microbial indices 41
    - 5.4.2. Bacterial and fungal contribution to microbial tissue 42
    - 5.4.3. Digestibility affected by N deficient diet 43
  - 5.5 Conclusions 44
  - 5.6 References 45

6. Effect of cattle faeces with different microbial biomass content on soil properties, gaseous emissions and plant growth 51

    6.1. Introduction 52

    6.2. Material and methods 53

        6.2.1. Soil 53

        6.2.2. Faeces sampling and quality determination 53

        6.2.3. Incubation experiment 54

        6.2.4. Pot experiment 55

        6.2.5. Microbial biomass C and N 55

        6.2.6. Ergosterol analysis 56

        6.2.7. Amino sugar analysis 56

        6.2.8. Inorganic N 57

        6.2.9. Statistical analysis 57

    6.3. Results 58

        6.3.1. Differences in faeces characteristics 58

        6.3.2. Effects of faeces types on soil microorganisms and grass growth 60

    6.4. Discussion 62

        6.4.1. Microbial indices 62

        6.4.2. Bacterial and fungal contribution to microbial tissue 63

        6.4.3. Effects of feeding regime and faeces composition 64

        6.4.4. Conclusions 66

    6.5. References 66

7. Zusammenfassung 72

8. Summary 74

9. Schlussfolgerung und Ausblick 76

10. Literatur 78

11. Danksagung 82

Abbildungsverzeichnis

Abb. 1a, b Pilzhyphen und -fruchtkörper auf Rinderkot nach 14 Tagen bei 25 °C 6

Abb. 2 Mikrobieller Biomasse-C in Rinderkot bei verschiedenen Konservierungsmethoden und Fütterungen. TM = Trockenmasse, HL = Hochleistung, LL = Niederleistung, HF = Färsen, N_2 = schockgefroren in Flüssigstickstoff bei -210 °C, Gefriertruhe = langsam eingefroren bei -18 °C, n = 6 9

Abb. 3 Rinderkotkonservierung in Flüssigstickstoff 9

Abb. 4 (Fig. 1): (a) Concentrations of microbial biomass C and (b) microbial biomass N in sub-samples of heifer 2 using different extractants over a 192 h incubation of extracts at 25°C; bars indicate ± one standard error; n = 4. 17

Abb. 5 (Fig. 2): (a) Concentrations of microbial biomass C, microbial biomass N, the microbial biomass C/N ratio; (b) concentrations of ergosterol and the ergosterol to microbial biomass C ratio during a 28-day incubation using faeces sub-samples of heifer 2 at 25°C; bars indicate ± one standard error; n = 4. 18

Abb. 6 (Fig. 1): Linear relationship between crude protein and microbial biomass N in faeces of dairy cows in relation to the different treatments (r = 0.55, n = 30, P < 0.01). 39

Abb. 7 (Fig. 2): Linear relationship between fungal C and ergosterol in faeces of dairy cows in relation to the different treatments (r = 0.43, n = 30, P < 0.01). 39

Abb. 8 (Fig. 3): Linear relationship between bacterial muramic acid and total N in faeces of dairy cows in rent treatments (r = -0.57, n = 30, P < 0.01). 40

Abb. 9 (Fig. 1): N_2O emissions from different soil treatments during 14 days of incubation at 22°C; bars indicate ± one standard error; n = 6. 60

Abb. 10 (Fig. 2): Harvest of Italian ryegrass after growing for 62 days on different soil treatments. Plant dry weight er pot, plant nitrogen content and mineral nitrogen immobilised from the soil per pot; bars indicate ± one standard error; different letters indicate a significant difference (Tukey/Kramer, $P < 0.05$); n = 6. 62

Tabellenverzeichnis

Tabelle 1 ATP-Gehalt und daraus errechnete Biomasse-C in Rinderkot 8

Tabelle 2 (Table 1): Variability of pH, elemental composition and organic components in cattle faeces from five identically fed heifers 13

Tabelle 3 (Table 2): Variability of microbial indices in cattle faeces from five identically fed heifers 19

Tabelle 4 (Table 3): Variability of microbial indices in cattle faeces from five identically fed heifers 20

Tabelle 5 (Table 1): Feed intake of silage and concentrate in kg per day and diet composition ... 34

Tabelle 6 (Table 2): Elemental composition and organic components in the faeces of dairy cows fed an N deficient and an N balanced diet 37

Tabelle 7 (Table 3): Microbial biomass, amino sugar indices and microbial C in faeces of dairy cows fed an N deficient and an N balanced diet 38

Tabelle 8 (Table 4): Microbial ratios in faeces of dairy cows fed an N deficient and an N balanced diet 38

Tabelle 9 (Table 1): Elemental composition and organic components in different cattle feeding regimes and in cattle faeces 58

Tabelle 10 (Table 2): Microbial biomass indices, amino sugar indices and microbial C in cattle faeces from different feeding regimes 59

Tabelle 11 (Table 3): Microbial indices in cattle faeces from different feeding regimes 59

Tabelle 12 (Table 4): Microbial indices, CO_2 emissions and mineral nitrogen in different soil treatments after 14 days of incubation 60

Abkürzungsverzeichnis

ADF	Säure-Detergenz-Faser
ADL	Säure-Detergenz-Lignin
Al	Aluminium
ANOVA	Varianzanalyse
$BaCl_2$	Barriumdichlorid
C	Kohlenstoff
Ca	Kalzium
$CaCl_2$	Kalziumchlorid
$CHCl_3$	Chloroform
C_{mik} (C_{mic})	Mikrobieller Biomasse-Kohlenstoff
CO_2	Kohlendioxid
$CuSO_4$	Kupfersulfat
DFG	Deutsche Forschungsgemeinschaft
DMSO	Dimethylsulfoxid
DNA	Desoxyribonukleinsäure
Fe	Eisen
H_2O	Wasser
HCl	Salzsäure
HNO_3	Salpetersäure
HPLC	Hochleistungsflüssigkeitschromatographie
ICP-AES	Induktiv gekoppeltes Hochfrequenzplasma-Atomemissionsspektrometer
K	Kalium
K_2SO_4	Kaliumsulfat
k_{EC}, k_{EN}	Extrahierbarer Teil des mikrobiellen Biomasse-Kohlenstoffs/-stickstoffs
KOH	Kaliumhydroxid
M	Molar
Mg	Magnesium
Mn	Mangan
N	Stickstoff
N_2O	Distickstoffmonoxid, Lachgas
Na	Natrium
NaOH	Natronlauge
Na_3PO_4	Trinatriumphosphat

NDF	Neutral-Detergenz-Faser
NH_3	Ammoniak
NH_4^+	Ammonium
NIRS	Nahinfrarot-Spektroskopie
N_{mik} (N_{mic})	Mikrobieller Biomasse-Stickstoff
O_2	Sauerstoff
OPA	*ortho*-Phthalaldehyd
P	Phosphor
P	Wahrscheinlichkeit
PCA	Perchlorsäure
PLFA	Phosphorlipidfettsäuren
r	Korrelationskoeffizient
S	Schwefel
UDN	unverdauter Futterstickstoff
WHC	Wasserhaltekapazität
$ZnSO_4$	Zinksulfat

1. Einleitung

In der Landwirtschaft ist die Erhaltung und Förderung der Bodenfruchtbarkeit ein grundlegendes Ziel. Mikroorganismen regulieren die Bodenparameter und stellen somit eine wichtige Messgröße für die Bodenqualität dar. Hier spielt die Nährstoffverfügbarkeit für das mikrobielle und pflanzliche Wachstum eine bedeutende Rolle.

1.1 Steuerung des Nährstoffhaushalts in der Ökologischen Landwirtschaft

In der Ökologischen Landwirtschaft werden keine synthetischen Dünger und Pestizide eingesetzt. Ziel ist, Nährstoffverluste zu minimieren (Vandermeer 1995; Mäder et al., 2002) und den Energieeinsatz zu optimieren (Kalk und Hülsbergen, 1996; Mäder et al., 2002). Die energetische Bilanz von ökologisch bewirtschafteten Betrieben kann günstiger sein als die von konventionellen (Salzgeber und Lörcher, 1997). Im Hinblick auf den innerbetrieblichen Nährstoffkreislauf und die Bodenfruchtbarkeit kommt der Düngerqualität eine große Bedeutung zu (Verhoeven et al., 1998).

Das Verhältnis von anorganischem zu organisch gebundenem Stickstoff sowie die C/N-Relation in Wirtschaftsdünger spielen eine wichtige Rolle für die Nutzung durch die Pflanze bzw. die Mikroorganismen im Boden. Dieses Verhältnis ist wiederum stark von der Rationszusammensetzung abhängig (Van Bruchem et al., 2000). Auch Keimgehalt und Darmmilieu werden in hohem Maße von der Fütterung beeinflusst. Daraus resultieren unterschiedliche Ausscheidungs- und Bindungsverhältnisse von Stickstoff, die auch das Emissionspotential von Ammoniak aus dem Kot erheblich beeinflussen. Im Gegensatz zum mineralischen Stickstoff wird der organisch gebundene Stickstoff nur durch den mikrobiellen Abbau pflanzenverfügbar. Dadurch kann er von den Pflanzen langsamer, aber auch längerfristig aufgenommen werden. Der mineralische Teil des Stickstoffes im Kot liegt überwiegend als pflanzenverfügbares Ammonium vor.

Bei Wiederkäuern wird die Stickstoffumsetzung vom Verhältnis der Energieversorgung zum Stickstoffgehalt und der Abbaubarkeit des Proteins im Pansen bestimmt (Gruber 1990). Die Ausscheidung von organisch gebundenem Kot-N kann durch Erhöhung der Energiezufuhr bzw. durch einen erhöhten Anteil an Rohfaser in der Ration gesteigert werden. Allerdings setzt eine erhöhte Energiezufuhr durch leicht verdauliche Kohlenhydrate das C/N-Verhältnis herab. Aerobe Zersetzung im Boden und anaerobe Verdaulichkeit im Pansen sind mikrobielle Vorgänge, die auf enzymatischen Abbau beruhen. Diese Prozesse werden überwiegend von den Inhaltsstoffen des abzubauenden Pflanzenmaterials wie Stickstoff- und Fasergehalt bzw. der Art der Wirtschaftsdünger bestimmt.

1. Einleitung

Über den Anteil an leicht verdaulichen Kohlenhydraten in der Futterration wird die Mikroflora im Pansen und im Dickdarm verändert. Mastrinder, deren Ration durch Kraftfutterbeigaben einen hohen Anteil leicht verdaulicher Kohlenhydrate enthielten, zeigten einen um mehr als 1000-fach höheren Keimgehalt im Kot im Vergleich zu mit Heu gefütterten Tieren (Diez-Gonzales et al., 1998; Harmon et al., 1999). Bei einer mit Rohprotein angereicherten Futterration traten zudem vermehrt phytotoxische und phenolhaltige Substanzen sowie biogene Amine auf, die negative Wirkungen auf den tierischen Organismus und die Mikroorganismen im Boden haben können (Van Bruchem et al., 2000). Die Fütterung verändert also die Zusammensetzung der Mikroorganismengemeinschaft im Dickdarm von Wiederkäuern und hat erhebliche Auswirkungen auf die mikrobiologischen Prozesse während der Mistlagerung und auf die Umsetzungsprozesse im Boden. Damit bildet die Fütterung eine zentrale Schnittstelle zwischen futterbaulichen und rationsgestaltenden Aspekten auf der einen Seite sowie der mikrobiellen Aktivität im Boden und der Nährstoffverfügbarkeit für das Pflanzenwachstum auf der anderen Seite. In welchem Ausmaß und mit welchem Aufwand durch die Fütterung der Humus- und Nährstoffhaushalt von Böden gesteuert werden kann, ist weitgehend unbekannt. Bislang liegen keine umfassenden Untersuchungen zum Einfluss der Fütterung von Wiederkäuern auf das C/N-Verhältnis und die Mikroorganismen im Kot vor. Auch die Effekte auf den Boden nach Rinderkotdüngung sind kaum untersucht.

1.2 Biologische Indikatoren der Bodenfruchtbarkeit

Bodenorganismen sind für die Bodenfruchtbarkeit, d.h. die dauerhafte Produktivität von landwirtschaftlich genutzten Anbausystemen, von großer Bedeutung, vor allem aufgrund ihrer Rolle als treibende Kraft für die Nährstoffversorgung von Pflanzen (Singh et al., 1989). Die mikrobielle Biomasse kann mit einigen Methoden relativ leicht und schnell ermittelt werden. Diese Methoden beruhen auf der Messung von organismenspezifischen Zellinhaltsstoffen (Jenkinson 1988), Glucose-aktivierter Atmung (Anderson und Domsch, 1978) oder der Extraktion von Kohlenstoff und Stickstoff aus mikrobiellen Zellen kurz nach deren Abtöten (Brookes et al., 1985; Vance et al. 1987).

Landwirtschaftlich genutzte Böden haben eine Biomasse zwischen 5 und 50 Tonnen Frischmasse pro Hektar (Jörgensen 1995). Die mikrobielle Biomasse wird besonders durch die Menge, aber auch durch die Qualität der eingetragenen organischer Substanz beeinflusst. Je größer der Eintrag, desto größer ist die mikrobielle Biomasse (Höper und Kleefisch, 2001). Diese zählt damit als wichtigster Indikator für Bodenfruchtbarkeit, der im Boden direkt gemessen werden kann. Neben dem Eintrag an organischer Substanz wird die mikrobielle Biomasse auch durch Klimafaktoren und verschiedene Bodeneigenschaften beeinflusst. Wie die einzelnen Einflussgrößen zusammenspielen, ist Gegenstand vielfältigster Forschungsbemühungen.

1.3 N_2O-Emissionen aus der Landwirtschaft

Ein wichtiger ökologischer Indikator zur Bewertung von Landbausystemen und ihrem Nährstoffmanagement ist die Belastung der Atmosphäre durch klimarelevante Emissionen. Die Freisetzung des Treibhausgases Distickstoffoxid (N_2O) aus landwirtschaftlich genutzten Böden ist hierbei von zentraler Bedeutung, da die Landwirtschaft der bedeutendste Verursacher anthropogener N_2O-Emissionen ist. Dieses Gas trägt sowohl zum globalen Treibhauseffekt als auch zum Ozonabbau in der Stratosphäre bei (IPCC, 1997). N_2O-Emissionen aus landwirtschaftlichen Produktionsflächen haben ihren Ursprung hauptsächlich in den Prozessen der Nitrifikation und Denitrifikation, die durch Bodenbakterien verursacht werden, (Firestone und Davidson, 1989). Die wichtigste Ursache erhöhter Emissionen aus landwirtschaftlich genutzten Böden ist der N-Eintrag. Ungenügend dokumentiert ist die Wirkung unterschiedlicher organischer Wirtschaftsdünger auf die Bildung und Emission von N_2O. Da die Datengrundlage nicht ausreicht, um Düngerabhängige Emissionsfaktoren anzugeben, wird als grobe Schätzung derzeit von einem einheitlichen Emissionsfaktor für organische und mineralische N-Düngung ausgegangen (1.25% des N-Eintrags unter Berücksichtigung von N-Verlusten als NH_3 und NO_x) (IPCC 2001). Unberücksichtigt bleibt hierbei, dass sich die N-Dynamik in Abhängigkeit der Düngerform erheblich unterscheiden kann. Bei organischen Düngern können N_2O-Emissionen durch eine gesteigerte O_2-Zehrung auch aus anderen N-Pools induziert werden (Flessa et al., 1995; Sehy 2004).

2. Ziele der Arbeit

Im Rahmen des DFG-Graduiertenkollegs "Steuerung des Humus und Nährstoffhaushalts in der Ökologischen Landwirtschaft" sollen im Fachgebiet Tierernährung und Tiergesundheit in Zusammenarbeit mit dem Fachgebiet Bodenbiologie und Pflanzenernährung die Auswirkungen der Futterration mit unterschiedlichen Anteilen leicht verdaulicher Kohlenhydrate in Kombination mit unterschiedlichen Rohprotein- und Rohfasergehalten, auf die mikrobielle Aktivität im Kot des Endarms sowie auf das C/N-Verhältnis im Kot untersucht werden. Bei dem zu bearbeitenden Projekt handelt es sich um ein interdisziplinäres Forschungsvorhaben an der Schnittstelle zwischen Bodenbiologie, Umweltchemie und Tierernährung.

Erforscht werden Interaktionen zwischen verschiedenen Nutzungsoptionen bei der Fütterung von Wiederkäuern und der Zielgröße Bodenfruchtbarkeit. Ziel ist die Beeinflussung der mikrobiellen Aktivität im Boden durch unterschiedliche Kotqualitäten zu erfassen und schließlich die Düngerqualität durch gezielte Fütterung zu steuern. Eine Kombination von unterschiedlichen Futterkomponenten soll zur Aufwertung des Wirtschaftsdüngers und zur Bodenverbesserung beitragen. Außerdem soll die Grundlage geschaffen werden, Ressourcennutzung effizienter zu gestalten und tierbedingte Umweltbelastungen zu verringern.

Die vielfältigen Interaktionen zwischen der Nährstoffzusammensetzung und den mikrobiellen Umsetzungen im Pansen und im Dickdarm der Wiederkäuer haben eine große Variation in der Beschaffenheit des Wirtschaftsdüngers zur Folge. Daraus resultiert ein entsprechendes Optimierungspotential im Hinblick auf eine zielgerichtete Abstimmung der beteiligten Steuerungsfaktoren mit dem Ziel, die Nutzung vorhandener Nährstoffpotentiale zu verbessern. Hier setzt das Forschungsvorhaben an. Gleichzeitig soll die Variationsbreite ermittelt werden, welche die Eigenschaften der Wirtschaftsdünger unter spezifischen Voraussetzungen beeinflusst. Diese führt zu einer variablen Düngungswirkung auf Bodenmikroorganismen und Pflanzen.

In Laborexperimenten soll die mikrobielle Biomasse aus Kotproben von Milchkühen ermittelt werden. Einen weiteren Schwerpunkt der Arbeiten bildet der Vergleich verschiedener Methoden zur Bestimmung mikrobieller Parameter in Kotproben. Die Untersuchungen schließen die Bestimmung der Nährstoffgehalte und Vergleiche zwischen Tieren verschiedener Alters- bzw. Nutzungsklassen ein. In einem Fütterungs-versuch sollen schließlich die Auswirkungen der Zusammensetzung von Futterrationen auf die Kotqualität untersucht werden. Die Kotproben werden für nachfolgende Untersuchungen zur Bestimmung der Mikroorganismenaktivität im Boden in Inkubationsexperimenten und zur Pflanzenverfügbarkeit von Stickstoff in Gewächshausgefäßversuchen verwendet. Ziel ist hierbei auch die Bewertung unterschiedlicher Fütterungsstrategien hinsichtlich der N-Umsetzung und der Emission von N_2O und CO_2.

3. Methodik und Vorversuche

Die nachfolgenden Methoden wurden an Rinderkot getestet und angepasst. Ausführliche Informationen zur Probenbehandlung finden sich in der ersten Publikation (Kapitel 4.2). Bisher nicht verwendete Ergebnisse, z.b. zur Methode der Adenylatbestimmung, werden ebenfalls dargestellt.

3.1 Chloroform-Fumigations-Extraktionsmethode (CFE)

Für Bodenproben wird zur Bestimmung der mikrobiellen Biomasse häufig die Chloroform-Fumigations-Extraktionsmethode (CFE) nach Vance et al. (1987) angewendet. Die Begasung (Fumigation) von Bodenmaterial mit $CHCl_3$ bewirkt eine Zerstörung der Zellmembranen lebender Organismen. Dadurch steigt nach der Entfernung des Chloroforms der Anteil an extrahierbaren Substanzen gegenüber einer unbegasten Bezugsbodenprobe (Jenkinson, 1966). Das Prinzip der CFE-Methode beruht darauf, dass diese Zunahme proportional zu der mikrobiellen Biomasse ist, die im Boden vorhanden war (Brookes et al., 1985; Vance et al., 1987).

Diese Vorgehensweise wurde an Kotproben in diesem Forschungsprojekt getestet, war jedoch in ihrer ursprünglichen Form in der Praxis nicht anwendbar, da sie keine messbaren Ergebnisse erzielte. Vollkommen andere Eigenschaften in der Textur, den Inhaltsstoffen und der Anzahl lebender bzw. toter Mikroorganismen verlangten einige Modifikationen der Methode (siehe 4.2.2.). Der Vorteil dieser Methode besteht vor allem in der ausschließlichen Erfassung intakter Zellen. So kann neben der Bakterien- und Pilzbiomasse auch die der im Kot vorkommenden Archeen quantifiziert werden.

3.2 Ergosterolbestimmung

Ergosterol ist ein wichtiger Bestandteil pilzlicher Zellmembranen (Weete und Weber, 1980) und wird daher als Biomarker für die Biomasse von Pilzen eingesetzt. Sterole machen bis zu 1% der Trockensubstanz der Pilze aus, wobei Ergosterol bei der Mehrzahl der Pilze einen Anteil von 90% und mehr einnehmen kann (Djajakirana et al., 1996).

Der Einfluss der Fütterung wird auch im Hinblick auf die pilzliche Biomasse (Abb. 1) untersucht und als vergleichende Größe der Ergosterolgehalt bestimmt. Die Methode von Djajakirana et al. (1996) bzw. Joergensen et al. (2000) mit Ethanolextraktion lieferte keine Messergebnisse an der HPLC. Auch verschiedene Extraktionsverhältnisse (1:200, 1:100, 1:50) sowie Zentrifugation der Extrakte führten zu keiner Verbesserung. Durch Verunreinigungen der Extrakte zeigten die Chromatogramme keine integrierbaren Ergosterolpeaks. Die Extraktion mit Petrolether nach Zelles et al. (1987) hingegen führte zum Erfolg und wurde für Rinderkot adaptiert. (Kapitel 4.2.3.)

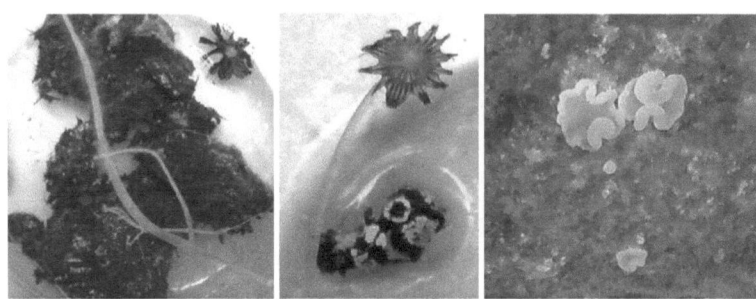

Abb. 1a, b Pilzhyphen und -fruchtkörper auf Rinderkot nach 14 Tagen bei 25 °C

3.3 Aminozuckerbestimmung

Im Boden gelten Aminozucker als mikrobielle Residuen (Amelung, 2001; Amelung et al., 2008). Muraminsäure ist ein Zellwandbestandteil von Bakterien. Glucosamin ist ebenfalls ein Bestandteil der Bakterienzellwand, daneben kommt es in den Zellwänden vieler Pilze vor. Aus dem Verhältnis von Muraminsäure und Glucosamin lässt sich das Bakterien-Pilzverhältnis bestimmen (Engelking et al., 2007). Die Herkunft von Galactosamin ist nicht geklärt, es wird aber angenommen, dass das meiste Galactosamin im Boden aus Bakterien stammt. Über die Herkunft von Mannosamin ist wenig bekannt.

Die Probenaufbereitung für die Aminozuckermessung mittels HPLC erfolgte zunächst nach Zhang und Amelung (1996) und schließlich nach Appuhn et al. (2004). Die erste Methode enthielt mehrere Aufreinigungsschritte und sollte für möglichst reine, leicht messbare Probenextrakte garantieren. Dieser Effekt war im Vergleich zur zweiten Methode jedoch gering. Zudem ergab Letztere höhere Aminozuckergehalte und erwies sich als anwenderfreundlicher, deshalb wurde sie für die Kotproben favorisiert (Kapitel 4.2.4).

3.4 Luminometrische ATP-Bestimmung

Adenosintriphosphat (ATP) ist eine wichtige Energiekomponente im Metabolismus aller Lebewesen (Wolstrup und Jensen, 2008). Der ATP-Gehalt des Bodens steht in engem Zusammenhang zu anderen Biomasseindices, wie z.b. mikrobieller Kohlenstoff, Stickstoff, usw. Er kann als unabhängiger Schätzwert des Biomassegehaltes im Boden dienen und zeigt einen starken Zusammenhang mit der CFE-Methode (Brookes et al., 1987). Probleme können bei der ATP-Extraktion auftreten, zum Beispiel eine unzureichende ATP-Freisetzung aus lebenden Zellen und ATP-Hydrolyse durch ATPasen. Verschiedene Extraktionsmethoden sollen für Rinderkot getestet und der ATP-Gehalt anschließend enzymatisch mit einem Luminometer gemessen werden (Luciferin-Luciferase-Assay nach Jenkinson und Oades, 1979).

Für die enzymatische Messung der Adenylatgehalte im Kot wurden verschiedene Extraktionsmethoden (Jenkinson und Oades, 1979; Brookes et al., 1987; Redmile-Gordon, 2011) und -verhältnisse getestet. Außerdem wurde als Vergleich die Methode von Dyckmans und Raubuch (1997) angewendet und an der HPLC gemessen. Die Extraktionseffizienz von Ultraschallsonde und Ultraschallbad wurde ebenfalls verglichen. Es ergaben sich keine signifikanten Unterschiede.

Als Extraktionsmittel wurden verwendet:
- Na_3PO_4/DMSO (Dyckmans und Raubuch, 1997)
- TCA+P (Jenkinson und Oades, 1979; Brookes et al., 1987)
- TCA+P+Imidazol (Redmile-Gordon, 2011)
- PCA (Jensen and Jørgensen, 1994)

Folgende Extraktionsverhältnisse wurden getestet (g Frischmasse:ml Extraktionsmittel):
DMSO: 1:10, 1:40, 1:80, 1:160
TCA: 1:12,5; 1:25, 1:50, 1:100, 1:150, 1:300, 1:400, 1:500, 1:1000, 1:5000

Höhere Extraktionsverhältnisse ergaben zwar höhere ATP-Werte, allerdings waren über einem Verhältnis von 1:50 die Standardabweichungen zu groß.

Um einen Fehler durch die Extraktfärbung der Proben auszuschließen, wurde schließlich für die Kontrollen (Blanks) und Standards der Puffer mit sterilisiertem Kot angesetzt, („Braune Matrix"). Stabile Ergebnisse mit geringer Streuung ergaben sich erst bei einer Vorinkubation der Extrakte nach Enzymzugabe von 1-2 Stunden auf Eis (Tabelle 1). Ohne Inkubation wurden keine stabilen Messwerte erreicht. Durch die chemische Lichtreaktion ergaben sich hier starke Variationen.

Tabelle 1 ATP-Gehalt und daraus errechnete Biomasse-C in Rinderkot

	PCA (1:20)	TCA + Imidazol (1:50)	TCA + Imidazol (1:100)
	(μg ATP g^{-1} TM)		
MW	5.3	2.3	6.2
CV (%)	19	14	33
Wiederfindung Probe (%)	83	68	102
Wiederfindung B-Blank (%)	74	98	105
Biomasse C* (μg g^{-1} TM)	636	276	744

MW = Mittelwert, TM = Trockenmasse, Färse 113 (PCA: n = 3, TCA: n = 5), *berechnet nach Oades und Jenkinson, 1979

Redmile-Gordon et al. (2011) wiesen im Boden ATP-Gehalte von 0.6-7.6 μg g^{-1} TM nach. Im Dickdarminhalt von Schweinen finden Jensen und Jørgensen (1994) 28 und Lindecrona et al., (2003) 14 μg ATP g^{-1} TM. Wenn man den Umrechnungsfaktor von Oades und Jenkinson (1979) annimmt und die ATP-Gehalte mit 120 multipliziert, so sind die errechneten Biomassewerte deutlich zu niedrig für Rinderkot, denn der mikrobielle Biomassegehalt aus der CFE-Methode beträgt 9.7 mg g^{-1} TM.

Ein Grund für die niedrigen ATP-Werte kann die Kotkonservierung durch das Einfrieren der Proben bei -18 °C sein. In Stickstoff eingefrorene Kotproben wurden noch nicht getestet.

3.5 Probenkonservierung

Im Zusammenhang mit den Experimenten für die dritte Publikation wurden zwei Methoden der Kotkonservierung verglichen. Die Kotproben wurden in zwei gleich große Portionen aufgeteilt, die eine Hälfte langsam bei -18 °C einfroren und die andere in flüssigem Stickstoff bei -210 °C schockgefroren. Die zweite Methode ergab deutlich geringere Standardabweichungen und einen höheren Gehalt an mikrobiellem Biomasse-C (Abb. 2). Teilweise erwiesen sich C- oder N-Gehalte in den Extrakten der Gefrier-proben (-18 °C) als nicht messbar. Aufgrund dieser Erkenntnisse wurden für die folgenden Experimente nur die Stickstoffproben verwendet.

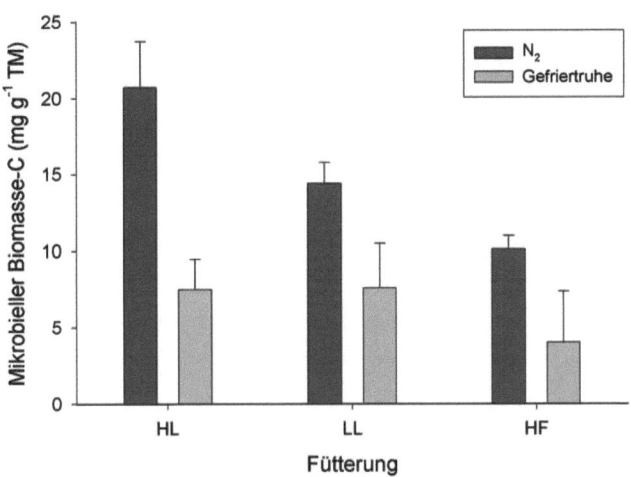

Abb. 2 Mikrobieller Biomasse-C in Rinderkot bei verschiedenen Konservierungsmethoden und Fütterungen. TM = Trockenmasse, HL = Hochleistung, LL = Niederleistung, HF = Färsen, N_2 = schockgefroren in Flüssigstickstoff bei -210°C, Gefriertruhe = langsam eingefroren bei -18°C, n = 6

Die Kotkonservierung mit Flüssigstickstoff erfolgt in handelsüblichen Gefrierbeuteln. Die homogenisierten Proben werden darin eingewogen, offen in eine Styroporkiste gelegt und der Stickstoff direkt auf die Proben gegossen. Die Kotproben sollten im Stickstoff verbleiben, bis sie vollständig durchgefroren sind (Abb. 3).

Abb. 3 Rinderkotkonservierung in Flüssigstickstoff

4. Determination of microbial biomass and fungal and bacterial distribution in cattle faeces

Soil Biology and Biochemistry (2011)

Daphne Isabel Jost [ab*], Caroline Indorf [b], Rainer Georg Joergensen [b], Albert Sundrum [a]

[a] Department of Animal Nutrition and Animal Health, University of Kassel, Nordbahnhofstr. 1a, 37213 Witzenhausen, Germany
[b] Department of Soil Biology and Plant Nutrition, University of Kassel, Nordbahnhofstr. 1a, 37213 Witzenhausen, Germany

Abstract

As an important component of organic fertilizers, animal faeces require methods for determining diet effects on their microbial quality to improve nutrient use efficiency in soil and to decrease gaseous greenhouse emissions to the environment. The objectives of the present study were (i) to apply the chloroform fumigation extraction (CFE) method for determining microbial biomass in cattle faeces, (ii) to determine the fungal cell-membrane component ergosterol, and (iii) to measure the cell-wall components fungal glucosamine and bacterial muramic acid as indices for the microbial community structure. Additionally, ergosterol and amino sugar data provide independent control values for the reliability of the microbial biomass range obtained by the CFE method. A variety of extractant solutions were tested for the CFE method to obtain stable extracts and reproducible microbial biomass C and N values, leading to the replacement of the original 0.5 M K_2SO_4 extractant for 0.05 M $CuSO_4$. The plausibility of the data was assessed in a 28-day incubation study at 25 °C with cattle faeces of one heifer, where microbial biomass C and N were repeatedly measured together with ergosterol. Here, the microbial biomass indices showed dynamic characteristics and possible shifts in the microbial community. In faeces of five different heifers, the mean microbial biomass C/N ratio was 5.6, the mean microbial biomass to organic C ratio was 2.2%, and the mean ergosterol to microbial biomass C ratio was 1.1‰. Ergosterol and amino sugar analysis revealed a significant contribution of fungi, with a percentage of more than 40% to the

---

[*] Corresponding author. Tel.: + 49 5542 98 1523; e-mail: daphne.jost@arcor.de

microbial community. All three methods are expected to be suitable tools for analysing the quality of cattle faeces.

Keywords: Microbial biomass C and N, Ergosterol, Glucosamine, Muramic acid

4.1 Introduction

Animal faeces are an important source of nutrients (Lovell and Jarvis, 1996; Ma et al., 2007; Wachendorf and Joergensen, 2011) and an important component of organic fertilizers, such as slurry and farmyard manure (FYM). The positive effects of FYM on activity and biomass of soil microorganisms have been repeatedly evaluated in long-term experiments on different soil types (Mäder et al., 2002; Böhme et al., 2005; Elfstrand et al., 2007). Soils fertilized with FYM contain higher contents of soil organic C and microbial biomass C, but lower contents of saprotrophic fungi and exhibit lower rates of microbial activity than those receiving similar amounts of C as straw (Scheller and Joergensen, 2008; Heinze et al., 2010). These data suggest that strong differences in the quality of organic fertilizers have strong effects on soil microorganisms and decomposition processes. However, long-term FYM experiments are usually based on the application of cattle manure without considering the variation of its quality. There is evidence that the fate of animal faeces in soil, i.e. its degradation and entering into the nutrient cycles, depends not only on the animal species but to a high degree on the feeding strategy (van Vliet et al., 2007). The N concentration of ruminant faeces has been shown to decrease with diet digestibility (Kyvsgaard et al., 2000; Sørensen et al., 2003). These differences in the concentration of nutrients and organic components also affect the microbial community of the faeces (van Vliet et al., 2007).

Considerable data is available on pathogenic and coliform bacteria in manure (van Kessel et al., 2007), but information on naturally occurring microorganisms is limited (Frostegård et al., 1997; Gattinger et al., 2007; van Vliet et al., 2007). However, it is well-documented from experiments with other organic substrates that the substrate colonising microbial community has an important influence on further decomposition processes (Flessa et al., 2002) and directly adds significant amounts of microorganisms to the autochthonous soil microbial biomass (Rasul et al., 2008). Cattle faeces contain a highly dynamic community of bacteria, archaea and fungi that has yet to be sufficiently quantified by the methods currently available. Most probable number and plate count approaches discriminate archaea and other non-cultivable microorganisms, which contribute more than 80% to the total number of species (Ouwerkerk and Klieve, 2001). Direct microscopic approaches often severely underestimate fungi (Joergensen and Wichern, 2008). The extraction of DNA followed by analysis of its composition provides important information on the microbial community composition of faeces (van Vliet et al., 2007; Sekhavati et al., 2009). However, DNA data usually provides no information on the biomass, due to losses during extraction, due to unknown or highly variable concentrations within different microbial species (Leckie et al., 2004; Joergensen and Emmerling, 2006) or due to its occurrence in dead microorganisms (Pisz et al., 2007; Bae and Wuertz, 2009). ATP would be a good candidate for the differentiation of living and dead microorganisms in faecal samples (Wolstrup and Jensen, 2008). However, in the anaerobic or micro-aerobic environment of faeces samples, the AEC and thus the ATP concentration within the microbial biomass is more variable than in soil (Jenkinson, 1988; Dyckmans et al., 2006).

The chloroform fumigation extraction (CFE) method has been applied in a wide range of substrates over the past two decades and is able to estimate the biomass of all microorganisms, surrounded by an intact cell membrane, i.e. bacteria, fungi, and archaea. The integrity of cell membranes is destroyed by chloroform, and especially the cytoplasmic part of the microbial constituents is further degraded by enzymatic autolysis and transformed into extractable components. This enables the CFE method to differentiate accurately between living and dead microbial tissue. The CFE method has also been successfully used to determine microbial biomass in a variety of substrates high in organic matter, such as litter layers (Joergensen and Scheu, 1999), peat (Brake et al., 1999), compost samples (Gattinger et al., 2004), and also pig manure (Aira et al., 2006), but not cattle manure. The strong comminution of the diet by cattle, followed by fermentation processes in the rumen in combination with a high water content and a highly dispersible crude protein fraction results in a pasty structure of cattle faeces, which may cause severe problems for the CFE method.

The first objective of the present study was to test the applicability of the CFE method for determining microbial biomass in cattle faeces. Thus, different extractants were tested on faeces to obtain stable extracts and reproducible microbial biomass C and N values. The second objective was to determine ergosterol as an index for fungal biomass and as an independent control method for CFE microbial biomass in cattle faeces. Ergosterol is an important component of fungal cell membranes, responsible for their stability (Weete and Weber, 1980) and has been successfully determined in a variety of solid substrates such as soils (Joergensen and Wichern, 2008; Strickland and Rousk, 2010) and roots (Appuhn and Joergensen, 2006), but not in faeces. Microbial biomass C and N were repeatedly measured together with ergosterol over a period of 28 days to reflect the behaviour of faecal microbial biomass during cattle faeces ageing. The third objective was to measure fungal glucosamine and bacterial muramic acid as further independent control values for both, CFE microbial biomass and ergosterol data. In soil, most amino sugars are bound to soil organic matter as microbial residues (Amelung, 2001; Amelung et al., 2008). However, in the freshly and dynamically decaying content of the gut, a large percentage of glucosamine and muramic acid is still in the fungal and bacterial biomass and not transferred to the fraction of microbial residues. In contrast to CFE microbial biomass and ergosterol, glucosamine has previously been used in rumen fluid of cattle (Sekhavati et al., 2009) and in sheep faeces for determination of the fungal biomass (Rezaeian et al., 2004a, b), as a basis for analysing the fungal community structure.

4.2 Materials and methods

4.2.1. Faeces sampling, quality determination, and incubation

Faeces samples were taken from five cattle heifers (*Bos primigenius taurus*, var. German Holstein) from a cattle breeding farm, located in Lower Saxony. All heifers were fed with the same silage mix of grass, maize (*Zea mays* L.) and sugar beet (*Beta vulgaris* L.) leaves for investigating the variation between individuals. The samples were taken rectally, immediately homogenised and frozen at -18 °C. A sub-sample was dried for 72 h at 65 °C and finely ground for chemical analyses. Total C and N were determined using a Vario MAX (Elementar, Hanau, Germany) elemental analyser. Total P, S, Na, K, Mg, Ca, Mn, Fe, and Al were analysed after HNO_3-pressure digestion as described by Chander et al. (2008) by ICP-AES (Spectro Analytical Instruments, Kleve, Germany). Faeces pH was measured using a 1/1 water/faeces ratio. Crude lipid was determined as described by van Soest (1967) in dried faeces (2.5 g), which were hydrolysed for 1 h with 4 M HCl, before the lipids were extracted from the residue with hexane in a Soxhlet apparatus. Crude ash, ammonium and the other organic components such as acid detergent lignin, acid detergent fibre, neutral detergent fibre, and crude fibre were determined by near-infrared spectroscopy (FOSS 6500, Rellingen, Germany) as described by Papke and Sundrum (personal communication) after appropriate calibration and validation. Table 1 shows mean values derived from five different heifers, each analysed in triplicate. Mean organic C varied around 428 mg g^{-1} dry weight (DW) with a C/N ratio of 16. The coefficient of variation (CV) between the three analytical replicates and five heifers was usually below 5%. An exception was the NH_4^+ content between the five heifers.

Freshly thawed faeces sub-samples (0.5 g DW) of heifer 2 were incubated for 28 days at 25°C in the dark in 40 replicates. For the determination of microbial biomass C and N (final method, see below) and ergosterol four replicate faeces portions for each method were removed from the incubator at day 0 and after 7, 14, 21 and 28 days of incubation, respectively.

Table 1 Variability of pH, elemental composition and organic components in cattle faeces from five identically fed heifers

	Mean heifer	CV replicate (%)	CV heifer (±%)
pH (H_2O)	8.3	0.69	2.9
(mg g^{-1} DW)			
C	428	0.75	1.4
N	22	0.19	3.0
NH_4^+	1.4	≤ 0.3	14
C/N	16	0.77	4.2
P	8.5	2.1	2.1
S	3.5	2.5	2.5
Na	1.0	3.6	3.6
K	5.8	2.1	2.1
Ca	22	2.6	2.6
Mg	8.0	2.6	2.6
Mn	0.4	2.3	2.3
Fe	3.6	4.9	4.9
Al	4.5	3.1	3.1
Crude lipid	42	4.4	7.5
Crude fibre	268	≤ 0.3	3.2
Neutral detergent fibre	430	≤ 0.3	4.7
Acid detergent fibre	362	≤ 0.3	1.9
Acid detergent lignin	94	≤ 0.3	4.9

DW = dry weight, CV = pooled coefficient of variation between replicate analyses (n = 3 per heifer) and heifers (n = 5)

4.2.2. Microbial biomass C and N estimation by CFE

Microbial biomass C and N were estimated by the chloroform fumigation extraction method (Brookes et al., 1985, Vance et al., 1987), testing different modifications of the extractant, the ratio of extractant to dry weight, and of fumigation handling. First, the following 7 extractants were tested: (1) 0.05 M K_2SO_4, (2) 0.5 M K_2SO_4, (3) 0.125 M $CaCl_2$, (4) 0.01 M $CuSO_4$, (5) 0.05 M $CuSO_4$, (6) 0.01 M $ZnSO_4$ and (7) 0.05 M $ZnSO_4$. All extractants were tested with faeces sub-samples of heifer 2 in three replicates each. Extractants 4 to 7 were further tested for stability by incubating extracts of fumigated and non-fumigated sub-samples of heifer 2 at 25°C in the dark in four replicates each. Organic C and total N were measured in the extracts after 0, 24, 48, 96 and 192 h. The effects of direct $CHCl_3$ application as drops to the sample (Mueller et al., 1992) and variation of fumigation time (20 h, 24 h and 28 h) were analysed. None of these variations led to a significant effect. Finally, faeces from the five heifers were measured with six replicates each. All faeces samples were fumigated, extracted and measured for C and N as described below.

Two freshly thawed subsamples equivalent to 0.5 g oven-dry faeces were taken for the analysis. One sub-sample was fumigated at 25°C with ethanol-free $CHCl_3$, which was removed after 24 h. Fumigated and non-fumigated portions were extracted with 100 ml 0.05 M $CuSO_4$ for 30 min by horizontal shaking at 200 rev min^{-1}. Following centrifugation (2000 g for 10 min), the faeces extract was filtered (hw3, Sartorius Stedim Biotech, Göttingen, Germany). Organic C in the extracts was measured as CO_2 by infrared absorption after combustion at 850 °C using a Dimatoc 100 automatic analyser (Dimatec, Essen, Germany). Microbial biomass C was calculated as follows: microbial biomass C = E_C/k_{EC}, where E_C = (organic C extracted from fumigated faeces) – (organic C extracted from non-fumigated faeces) and k_{EC} = 0.45 (Wu et al., 1990; Joergensen, 1996). Total N in the extracts was measured by chemoluminescence detection. Microbial biomass N was calculated as follows: microbial biomass C = E_N/k_{EN}, where E_N = (organic N extracted from fumigated faeces) – (organic N extracted from non-fumigated faeces) and k_{EN} = 0.54 (Brookes et al., 1985; Joergensen and Mueller, 1996). C and N standards for calibration of the Dimatoc 100 analyser were prepared in 0.05 M $CuSO_4$ solution.

4.2.3. Ergosterol analysis

The extraction method of Zelles et al. (1987) was used for determination of the fungal cell-membrane ergosterol. Freshly thawed faeces (0.5 g DW) were placed into 30 ml test tubes and treated with 10 ml methanol, 2.5 ml ethanol and 1 g KOH. Each sample was saponified for 90 min at 70 °C under reflux. After cooling, ergosterol was extracted in two steps with 15 + 10 ml petroleum ether. From the supernatant, 15 ml were evaporated in a vacuum rotary evaporator at 40 °C. The non-polar fraction was dissolved in 5 ml methanol and stored at 4 °C until measurement. Ergosterol was determined by reversed-phase HPLC with 100% methanol as the mobile phase and detected at a wavelength of 282 nm.

4.2.4. Amino sugar analysis

The amino sugars muramic acid, glucosamine and galactosamine were determined with modifications according to Appuhn et al. (2004) as described by Indorf et al. (2011). Moist samples of 2 g fresh faeces were weighed into 20 ml test tubes, mixed with 10 ml 6 M HCl, and heated for 2 h at 105 °C. After HCl removal from the filtered hydrolysates in a vacuum rotary evaporator at 40 °C and centrifugation, the samples were transferred to vials and stored at -18 °C until the HPLC measurements. Chromatographic separations were performed on a Phenomenex (Aschaffenburg, Germany) Hyperclone C_{18} column (125 mm length × 4 mm diameter), protected by a Phenomenex

C_{18} security guard cartridge (4 mm length × 2 mm diameter) at 35°C. The HPLC system consisted of a Dionex (Germering, Germany) P 580 gradient pump, a Dionex Ultimate WPS – 3000TSL analytical autosampler with in-line split-loop injection and thermostat and a Dionex RF 2000 fluorescence detector set at 445 nm emission and 330 nm excitation wavelengths. For the automated pre-column derivatisation, 50 µl OPA and 30 µl sample were mixed in the preparation vial and after 120 sec reaction time15 µl of the indole derivates were injected. The mobile phase consisted of two eluents and was delivered at a flow rate of 1.5 ml min^{-1}. Eluent A was a 97.8/0.7/1.5 (v/v/v) mixture of an aqueous phase, methanol and tetrahydrofuran (THF). The aqueous phase contained 52 mmol sodium citrate and 4 mmol sodium acetate, adjusted to pH 5.3 with HCl. Then methanol and THF were added. Eluent B consisted of 50% water and 50% methanol (v/v).

Fungal C (mg g^{-1} dry weight) was calculated by subtracting bacterial glucosamine from total glucosamine as an index for fungal residues, assuming that muramic acid and glucosamine occur at a 1 to 2 molar ratio in bacterial cells (Engelking et al., 2007): mg fungal C g^{-1} DW = (mmol glucosamine – mmol muramic acid) × 179.2 g mol^{-1} × 9, where 179.2 is the molecular weight of glucosamine and 9 the conversion value of fungal glucosamine to fungal C (Appuhn and Joergensen, 2006). Bacterial C (µg g^{-1} DW) was calculated as an index for bacterial residues by multiplying the concentration of muramic acid in µg g^{-1} DW by 45 (Appuhn and Joergensen, 2006).

4.2.5. Statistical analysis

The results presented in the tables are arithmetic means and expressed on an oven-dry basis (about 72 h at 60 °C). The significance of difference was tested by one-way analysis of variance. All statistical analyses were performed using JMP 7.0 (SAS Inst. Inc.).

4.3 Results

4.3.1. Extractant efficiency, variability, and stability of the CFE method

Mostly no CHCl$_3$-labile N was extractable with 0.05 M K$_2$SO$_4$, 0.5 M K$_2$SO$_4$ or 0.125 M CaCl$_2$. Microbial biomass C using these extractants was 21, 20 and 15 mg g^{-1} DW, respectively (data not shown), with a mean coefficient of variation (CV) of 50%. Due to the problems described and the CV values, extractants 1 to 3 were not further tested. Mean microbial biomass C extracted with ZnSO$_4$ and CuSO$_4$ was 27 mg g^{-1} DW for the 0.01 M concentration and 31 mg g^{-1} for the 0.05 M concentration. Microbial biomass N varied around 4.5 mg g^{-1} for all four extractants. The mean CV for all four extractants ranged from 10 to 12% for microbial biomass C and N.

At the beginning of the incubation, the microbial biomass C ranged from 22 to 31 mg g^{-1} for all extractants (Fig. 1a) and the microbial biomass N concentration from 3.6 to 4.4 mg g^{-1} (Fig. 1b).

4. Determination of microbial biomass and fungal and bacterial distribution in cattle faeces

After 192 h, the microbial biomass C decreased to 4.5 and 11 mg g^{-1} in the 0.01 and 0.05 M ZnSO$_4$ extracts, respectively. In contrast, the C concentration of the CuSO$_4$ extracts remained constant. Similar characteristics were found for microbial biomass N. The extracts with CuSO$_4$ had CV values of 3.1 and 3.3%, for microbial biomass C and N, whereas the ZnSO$_4$ extracts had CV values between 26 and 57%. Differences between the 0.01 and 0.05 M CuSO$_4$ extracts were lower, at 4.5 and 2.0% respectively.

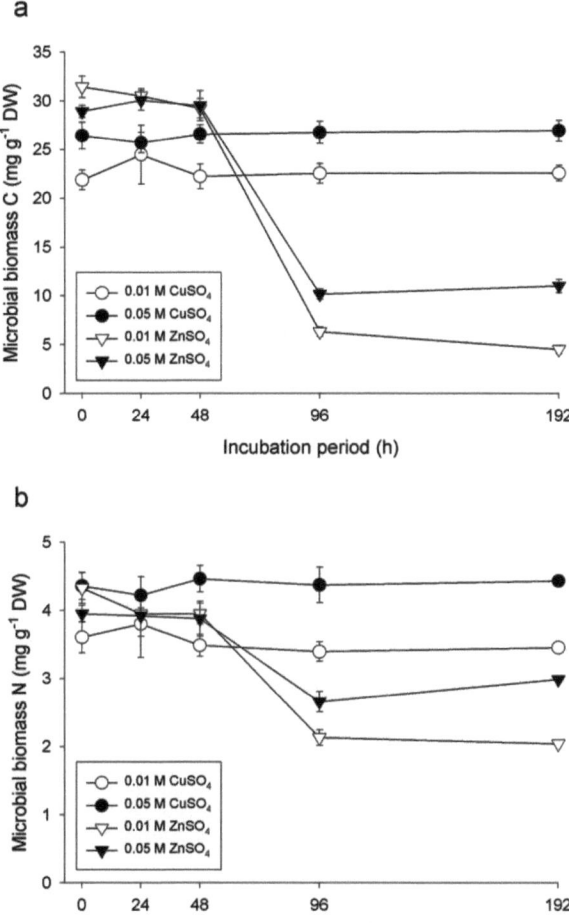

Fig. 1. (a) Concentrations of microbial biomass C and (b) microbial biomass N in sub-samples of heifer 2 using different extractants over a 192 h incubation of extracts at 25°C; bars indicate ± one standard error; n = 4.

4.3.2. Microbial biomass and ergosterol in incubated faeces

During the 28-day incubation study, the concentrations of microbial biomass C and N in the faeces of heifer 2 started at 7.4 and 1.9 mg g^{-1} DW, respectively, but only microbial biomass C nearly doubled within two days (Fig. 2a), also doubling the microbial biomass C/N ratio. Microbial biomass C declined to the initial concentrations and remained more o less stable as microbial biomass N until the end of the incubation. The ergosterol concentration declined by 60% within 3 days, increased roughly to the initial concentration at day 14 and declined again (Fig. 2b). The ergosterol to microbial biomass C ratio varied around a mean of 2.4‰, ranging from 0.8 to 5.0‰, and showed a strong negative correlation with the microbial biomass C/N ratio ($r = -0.81$, $P < 0.01$).

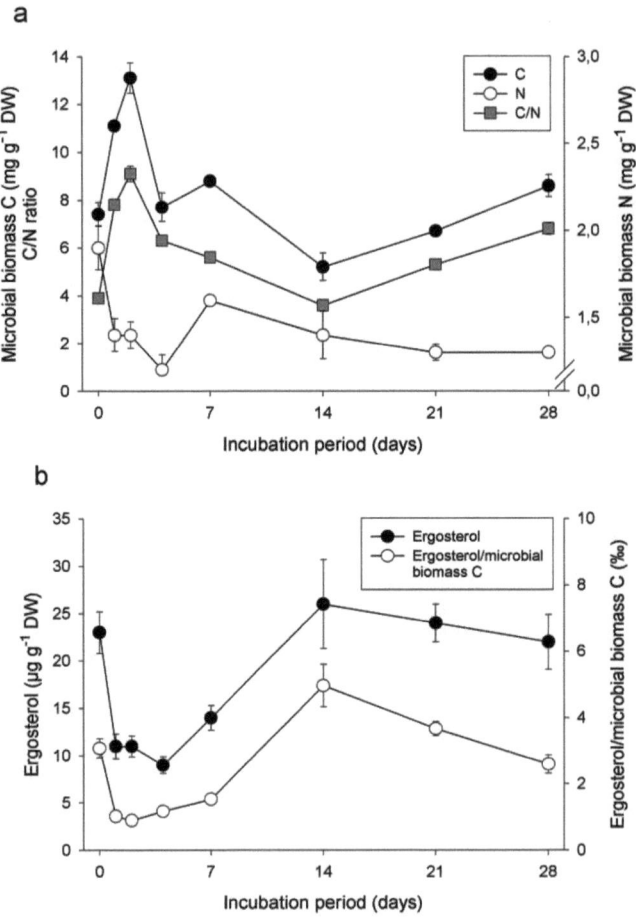

Fig. 2. (a) Concentrations of microbial biomass C, microbial biomass N, the microbial biomass C/N ratio; (b) concentrations of ergosterol and the ergosterol to microbial biomass C ratio during a 28-day incubation using faeces sub-samples of heifer 2 at 25°C; bars indicate ± one standard error; n = 4.

4.3.3. Relationships of microbial indices in faeces of different heifers

In faeces of five different heifers, the means of microbial biomass C and N were 9.3 and 1.9 mg g^{-1} DW, respectively (Table 2). The mean microbial biomass C/N ratio was 5.6. The mean ergosterol content was 9.4 µg g^{-1} DW and the resulting ergosterol to microbial biomass C ratio 1.1‰. Microbial biomass N and ergosterol were significantly correlated ($r = 0.93$, $n = 6$, $P < 0.05$). The mean concentrations of muramic acid, galactosamine and glucosamine were 0.43, 1.4 and 2.2 mg g^{-1} DW, respectively (Table 3). On the basis of these data, a mean concentration of microbial C of 33 mg g^{-1} DW, a mean fungal C to bacterial C ratio of 0.78 and mean fungal glucosamine to ergosterol ratio of 190 could be calculated. In contrast to muramic acid, glucosamine and galactosamine both showed significant negative correlations with ergosterol ($r = -0.97$, $P < 0.01$ and $r = -0.94$, $P < 0.05$, respectively).

Table 2

Variability of microbial indices in cattle faeces from five identically fed heifers

Heifer	Microbial biomass			Ergosterol	Ergosterol/ microbial biomass C
	C	N	C/N		
	(mg g^{-1} DW)			(µg g^{-1} DW)	(‰)
1	8.5 ab	2.4 a	4.8	13.0 a	1.6
2	9.7 ab	2.1 a	4.7	10.5 ab	1.1
3	4.9 b	1.7 a	3.0	7.4 b	1.3
4	9.5 ab	1.4 a	7.4	6.1 b	0.8
5	13.8 a	1.7 a	7.9	9.8 ab	0.9
Mean	9.3	1.9	5.6	9.4	1.1
CV (±%)	28	21		26	13

DW = dry weight; CV = pooled coefficient of variation between replicate sub-samples (n = 6); different letters within a column show significant differences (Tukey, $P = 0.05$)

Table 3 Variability of microbial indices in cattle faeces from five identically fed heifers

Heifer	MurN	GlcN	GalN	Microbial C	Fungal C/ bacterial C	Fungal GlcN/ ergosterol
			(mg g^{-1} DW)			
1	0.28 b	1.7 b	1.1 c	25 b	0.90	
2	0.47 a	2.2 ab	1.3 bc	35 a	0.64	
3	0.43 a	2.5 a	1.4 ab	36 a	0.88	
4	0.46 a	2.4 a	1.6 a	37 a	0.78	
5	0.42 a	2.1 ab	1.3 bc	33 ab	0.72	
Mean	0.43	2.2	1.4	33	0.78	
CV (±%)	16	15	12	15	10	

DW = dry weight; MurN = muramic acid; GlcN = glucosamine; GalN = galactosamine; CV = pooled coefficient of variation between replicate sub-samples (n = 6); different letters within a column show significant differences (Tukey, $P = 0.05$)

4.4 Discussion

4.4.1. Microbial biomass in cattle faeces by the CFE method

The microbial biomass C values of cattle faeces obtained by the CFE method are within the range of maximum values reported in other substrates high in organic matter, such as composted straw (7.3 mg C g^{-1} DW: Joergensen et al., 1997) and cattle manure compost (12.6 mg C g^{-1} DW: Gattinger et al., 2004). In non-composted pig slurry, Aira et al. (2006) and Guerrero et al. (2007) reported microbial biomass contents ranging between 4.2 and 8.9 mg C g^{-1} DW, using the k_{EC} values of 0.38 proposed by Vance et al. (1987) for dichromate oxidation. In this study, the mean microbial biomass C/N ratio close to 6 in cattle faeces is in the lower range of values reported for soils (Jenkinson, 1988; Joergensen and Mueller, 1996). The mean microbial biomass C to organic C ratio of 2.2% is within the range of different soils (Anderson and Domsch, 1989; Joergensen, 2010). These similarities give confidence that the results for cattle faeces obtained by the CFE method are common.

It is also possible to use the k_{EC} and k_{EN} values already applied to soil and other substrates for converting CHCl$_3$ labile C and N to microbial biomass C and N in faeces, respectively. These values were mainly calibrated indirectly against the chloroform fumigation incubation (CFI) method (Jenkinson, 1988; Wu et al., 1990). The large range of the k_{EC} and k_{EN} values reported in the literature and summarized by Joergensen (1996) and Joergensen and Mueller (1996), respectively, are often caused by the methodological limitations of the CFI method and not by true changes in the conversion values (Joergensen et al., 2011). However, it cannot be completely ruled out that such true changes may occur in the dynamic environment of animal faeces. Young cells contain more easily soluble components in the cytoplasm than older cells (Bremer and van Kessel, 1990), leading to increased conversion values. More $CHCl_3$ labile material has been rendered extractable by direct fumigation from large fungal than from small bacterial cells (Eberhardt et al., 1996), leading to decreased conversion values in bacteria dominated communities. However, it is certainly more appropriate to use a non-perfect conversion value, acceptable for several sample types, than not to use one at all, as fumigation *never* renders 100% of the microbial biomass extractable (Joergensen et al., 2011).

In cattle manure, between 300 nmol (Frostegård et al., 1997) and 1500 nmol (Gattinger et al., 2007) phospholipid fatty acids (PLFA) were detected. These PLFA contents were converted to microbial biomass C by multiplying them by 5.8 as suggested by Joergensen and Emmerling (2006), resulting in a range from 1.7 to 8.7 mg C g^{-1} DW, similar to the range obtained by the CFE method. However, it should be noted that this conversion value is based on a limited number of observations, again solely investigating soil microorganisms (Bailey et al., 2002). It has been shown that cattle manure contains 240 nmol phospholipid ether lipids (PLEL), which means that archaea contribute an additional 20% to the phospholipid pool (Gattinger et al., 2007). No information is available for conversion of PLEL to the biomass of archaea.

The agreement of the microbial biomass data from cattle faeces in this study with those obtained from soil and related organic components is extraordinary, considering the absence of any structure in the pasty cattle faeces, combined with a highly dispersible crude protein fraction. However, the 0.05 M $CuSO_4$ sufficiently flocculates the protein-rich faeces, which is inevitably necessary to extract $CHCl_3$ labile N components from the cattle faeces. The divalent Cu^{2+} ions have a stronger flocculation capacity than the monovalent K^+ ions in K_2SO_4. In addition, the Cu strongly inhibits microbial decomposition of the highly decomposable $CHCl_3$ labile organic material obtained after extraction, as demonstrated in the incubation experiment with extracts at 25°C. The main reason for the relatively high CV of the cattle faeces samples is, other than the variability mentioned within the test animals, most likely the small sample size with variations in the dry matter concentration, which has to be determined on separate sub-samples.

4.4.2. Bacterial and fungal distribution in cattle faeces

In situations in which the microbial colonisation of living or freshly decaying organic material is to be analysed, such as excised roots, glucosamine and muramic acid can be used as an index for fungal and bacterial biomass (Appuhn and Joergensen, 2006). In this experiment, by combining fumigation extraction and ergosterol data, Appuhn and Joergensen (2006) estimated an average microbial biomass C content of 11 mg g^{-1} DW in plant roots. In contrast, the mean microbial C was 23 mg g^{-1} DW, based on the concentrations of muramic acid and fungal glucosamine. In cattle faeces, the corresponding values were 9.3 mg microbial biomass C g^{-1} DW and 33 mg microbial C g^{-1} DW, respectively. This suggests that roughly 28% of the microbial tissue in the cattle faeces belongs to the living fraction and the other 72% are remains of dead fungi and bacteria. In the highly dynamic situation of C and N supply in the gut, rapid microbial growth is probably accompanied by concomitant microbial death similar to that observed in soil (Chander and Joergensen, 2001). For this reason, these figures seem to be realistic, but should not be stressed too much, because it is not known (1) whether the glucosamine and muramic acid concentrations are identical in living and dead microbial tissue and (2) whether the conversion values may lead to an underestimation of microbial biomass C.

In this study, an average fungal to bacterial C ratio of 0.78 corresponds to 44% fungal C and 56% bacterial C. This suggests that the faecal microbial community is less clearly dominated by bacteria than usually imagined (Frostegård et al., 1997; Griffith et al., 2009), therefore often disregarding fungi (van Vliet, 2007; Frey et al., 2010). However, this view neglects the significant contribution of archaea to the microbial biomass of cattle faeces, as archaea do not contain the amino sugar murein acid. In contrast to fumigation extraction, chitin, i.e. glucosamine, has been previously used in rumen fluid of cattle (Sekhavati et al., 2009) and in sheep faeces for the determination of anaerobic fungi (Rezaeian et al., 2004a, b). The chitin concentration in sheep faeces was 10.2 mg g^{-1} DW (Rezaeian et al., 2004a). Compared with the glucosamine concentration in cattle faeces, this chitin concentration was considerably higher. However, this technique cannot distinguish between fungal glucosamine, bacterial glucosamine, and galactosamine (Chen and Johnson, 1983). This certainly leads to overestimation of fungal tissue. The negative relationship between galactosamine and ergosterol suggests that galactosamine was mainly of bacterial origin, in accordance with the general view in soil science (Amelung et al., 2008), although fungi also produce galactosamine (Engelking et al., 2007).

The ergosterol to microbial biomass C ratio has been repeatedly used as an index for the contribution of fungi to the total microbial biomass (Bååth and Anderson, 2003). Of the anaerobic fungal populations, yeasts such as *Candida* sp. (Ahmad et al., 2010) or *Saccharomyces cerevisia* (Aguilera et al., 2006) contain high concentrations of ergosterol, also food spoiling *Mucor plumbeus* (Taniwaki et al., 2009). In contrast, no ergosterol but high concentrations of cholesterol were measured in chytridiomycetes (Weete et al., 1989; Kagami et al., 2007). Until now, no information is available on the ergosterol concentration of anaerobic fungal species found in cattle such *Anaeromyces*, *Orpinomyces*, *Caecomyces*, or *Piromyces* (Griffith et al., 2009). The mean ratio of ergosterol to microbial biomass C of 1.1‰ found in cattle faeces is markedly below that of soils (Joergensen and Wichern, 2008). Klamer and Bååth (2004) obtained a factor of 190 for converting ergosterol to fungal biomass C in 24 compost fungi, which is within the range presented by Joergensen and Wichern (2008) and considerably above the factor of 90 reported by Djajakirana et al. (1996). Using the factor reported by Klamer and Bååth (2004) would result in a mean fungal biomass C content of 1.8 mg C g^{-1} DW in the faeces samples in this study. However, not all faecal fungi may contain ergosterol as stated above, but all contain glucosamine. The mean ratio of fungal glucosamine to ergosterol was 190 in cattle faeces and thus somewhat above the mean ratio observed by Appuhn and Joergensen (2006) in plant roots. During the 4-week incubation of cattle faeces performed in this study, a decrease in the ergosterol to microbial biomass ratio was followed by an increase, which may indicate a shift in the microbial community, especially fungal communities to increasing aerobiosis and decreasing C availability. This estimation is supported by the inverse changes in the microbial biomass C/N ratio over the incubation period, i.e. an increase was followed by a decrease. However, in dynamic systems, ergosterol tends to accumulate for certain periods after fungal death (Zhao et al., 2005), which should also be considered in the present case.

The majority of microbial data in faeces samples is based on colony forming units (CFU) or thallus forming units (TFU), counted on plates or as the most probable number (Griffith et al., 2009). Although the unknown representation of cultivable microorganisms has raised doubt about converting these counts into microbial biomass (Ritz, 2007), the data of this study are also within the huge range produced by these approaches. Between 1.4 and 14 mg C g^{-1} DW coliform bacteria, mainly *Escherichia coli* have been found in cattle faeces (Avery et al., 2004; van Kessel et al., 2007), assuming a mass of 6.25 x 10^{-10} g per single *E. coli* bacterium (Siu, 2003) and a C concentration of 45% in microbial cells (Jenkinson, 1988). In agreement with these CFU data, van Vliet et al. (2007) observed between 8.6 and 10.9 x 10^{-10} g per single bacterial cell in cattle faeces, i.e. 40 to 70% higher values on the basis of confocal laser scanning microscopy using europium chelate/fluorescent brightener differential stain in combination with automated image analysis (Bloem et al., 1994). The bacterial biomass ranged from 1.2 to 8.0 mg C g^{-1} DW, depending on the quality of the diet (van Vliet et al., 2007). Nevertheless, it should be considered that fungal biomass is not included in this technique.

Roughly 3.3 mg g^{-1} DW fungal biomass has been found in cattle faeces (Griffith et al., 2009), assuming a mass of 3.3 ng per single fungal count, using the geometric mean of a range from 0.4 to 52 ng per fungal count (Schnürer, 1993). According to Frostegård et al. (1997), cattle manure contains 2.6 mol% 18:2ω6,9 (linoleic acid) and 5.3 mol% 18:1ω9 (oleic acid), which is similar to mean values found in grassland soils on the basis of 508 and 88 observations, respectively (Joergensen and Wichern, 2008). If linoleic acid is converted to fungal biomass by multiplying by 107 (Joergensen and Wichern, 2008), total PLFA contents of 300 nmol (Frostegård et al., 1997) and

1500 nmol (Gattinger et al., 2007) correspond to a fungal biomass C ranging from 0.8 to 4.2 mg g^{-1} DW. It could be concluded from these values that fungi contribute roughly 48% to the microbial biomass and bacteria to 52%, again neglecting archaea, i.e. values nearly identical to those based on the amino sugar analysis.

4.4.3. Conclusions

This study demonstrated the applicability of the CFE method for reliable determination of microbial biomass C and N in cattle faeces. The replacement of 0.5 M K_2SO_4 by 0.05 M $CuSO_4$ as the extractant solved previous difficulties in extracting $CHCl_3$ labile N components and ensured stable faeces extracts. In combination with the fungal ergosterol, the CFE method indicated a shift in the microbial community structure during ageing of cattle faeces. Ergosterol and amino sugar analysis revealed a significant contribution of fungi, with a biomass percentage of more than 40% to the microbial community. The CFE method and amino sugar analysis gave results on a similar level, although fungal glucosamine and bacterial muramic acid revealed a considerable presence of dead microbial tissue in faeces. All three methods are expected to be suitable tools for analysing the effects of diet composition on faeces quality and the subsequent behaviour of animal manure in soil.

Acknowledgments

The skilful technical assistance of Gabriele Dormann and Christiane Jatsch is highly appreciated. We also thank Mick Locke for carefully correcting our English. This project was supported by a grant of the Research Training Group 1397 "Regulation of soil organic matter and nutrient turnover in organic agriculture" of the German Research Foundation (DFG).

4.5 References

Aguilera, F., Peinado, R.A., Millán, C., Ortega, J.M., Mauricio, J.C., 2006. Relationship between ethanol tolerance, H$^+$-ATPase activity and the lipid composition of the plasma membrane in different wine yeast strains. International Journal of Food Microbiology 110, 34-42.

Ahmad, A., Khan, A., Manzoor, N., Khan, L.A., 2010. Evolution of ergosterol biosynthesis inhibitors as fungicidal against *Candida*. Microbial Pathogenesis 48, 35-41.

Aira, M., Monroy, F., Domínguez, J., 2006. Changes in microbial biomass and microbial activity of pig slurry after the transit through the gut of the earthworm *Eudrilus eugeniae*. Biology and Fertility of Soils 42, 371–376.

Amelung, W., 2001. Methods using amino sugars as markers for microbial residues in soil. In: Lal, J.M., Follett, R.F., Stewart, B.A. (Eds.), Assessment Methods for Soil Carbon. Lewis Publishers, Boca Raton, pp. 233-272.

Amelung, W., Brodowski, S., Sandhage-Hofmann, A., Bol, R., 2008. Combining biomarker with stable isotope analyses for assessing the transformation and turnover of soil organic matter. Advance in Agronomy 100, 155-250.

Anderson, T.-H., Domsch, K.H., 1989. Ratios of microbial biomass carbon to total organic carbon in arable soils. Soil Biology & Biochemistry 21, 471-479.

Appuhn, A., Joergensen, R.G., 2006. Microbial colonisation of roots as a function of plant species. Soil Biology & Biochemistry 38, 1040-1051.

Appuhn, A., Joergensen, R.G., Raubuch, M., Scheller, E., Wilke, B., 2004. The automated determination of glucosamine, galactosamine, muramic acid and mannosamine in soil and root hydrolysates by HPLC. Journal of Plant Nutrition and Soil Science 167, 17-21.

Avery, S.M., Moore, A., Hutchison, M.L., 2004. Fate of *Escherichia coli* originating from livestock faeces deposited directly onto pasture. Letters in Applied Microbiology 38, 355–359.

Bååth, E., Anderson, T.H., 2003. Comparison of soil fungal/bacterial ratios in a pH gradient using physiological and PLFA-based techniques. Soil Biology & Biochemistry 35, 955-963.

Bae, S., Wuertz, S., 2009. Discrimination of viable and dead fecal bacteroidales bacteria by quantitative PCR with propidium monoazide. Applied and Environmental Microbiology 75, 2940-2944.

Bailey, V.L., Peacock, A.D., Smith, J.L., Bolton, H. JR., 2002. Relationships between soil microbial biomass determined by chloroform fumigation–extraction, substrate-induced respiration, and phospholipid fatty acid analysis. Soil Biology & Biochemistry 34, 1385–1389.

Bloem, J., Lebbink, G., Zwart, K.B., Bouwman, L.A., Burgers, S.L., de Vos J.A., de Ruiter, P.C., 1994. Dynamics of microorganisms, microbivores and nitrogen mineralisation in winter wheat fields under conventional and integrated management. Agriculture, Ecosystems and Environment 51, 129-143.

Böhme, L., Langer, U., Böhme, F., 2005. Microbial biomass, enzyme activities and microbial community structure in two European long-term field experiments. Agriculture Ecosystems and Environment 109, 141-152.

Brake, M., Höper, H., Joergensen, R.G., 1999. Land use-induced changes in activity and biomass of microorganisms in raised bog peats at different depths. Soil Biology & Biochemistry 31, 1489-1497.

Bremer, E., van Kessel, C., 1990. Extractability of microbial ^{15}C and ^{15}N following addition of variable rates of labelled glucose and $(NH_4)_2SO_4$ to soil. Soil Biology & Biochemistry 22, 707-713.

Brookes, P.C., Landman, A., Pruden, G., Jenkinson, D.S., 1985. Chloroform fumigation and the release of soil nitrogen: a rapid direct extraction method for measuring microbial biomass nitrogen in soil. Soil Biology & Biochemistry 17, 837-842.

Chander, K., Hartmann, G., Joergensen, R.G., Khan, K.S., Lamersdorf, N., 2008. Comparison of three methods for measuring heavy metals in soils contaminated by different sources. Archives of Agronomy and Soil Science 54, 413-422.

Chander, K., Joergensen, R.G., 2001. Decomposition of ^{14}C glucose in two soils with different levels of heavy metal contamination. Soil Biology & Biochemistry 33, 1811-1816.

Chen, G.C., Johnson, B.R., 1983. Improved colorimetric determination of cell wall chitin in wood decay fungi. Applied and Environmental Microbiology 46. 13–16.

Djajakirana, G., Joergensen, R., Meyer, B., 1996. Ergosterol and microbial biomass relationship in soil. Biology and Fertility of Soils 22, 299–304.

Dyckmans, J., Flessa, H., Lipski, A., Potthoff, M., Beese, F., 2006. Microbial biomass and activity under oxic and anoxic conditions as affected by nitrate additions. Journal of Plant Nutrition and Soil Science 169, 108-115.

Eberhardt, U., Apel, G., Joergensen, R.G., 1996. Effects of direct chloroform-fumigation on suspended cells of ^{14}C and ^{32}P labelled bacteria and fungi. Soil Biology & Biochemistry 28, 677-679.

Edwards, J., Huws, S., Kingston-Smith, A., Jimenez, H., Skøt, K., Griffith, G.W., McEwan, N.R., Theodorou, M.K., 2008. Dynamics of initial colonisation of non-conserved perennial ryegrass by anaerobic fungi in the bovine rumen. FEMS Microbiology Ecology 66, 537–546.

Elfstrand, S., Hedlund, K., Mårtensson, A., 2007. Soil enzyme activities, microbial community composition and function after 47 years of continuous green manuring. Applied Soil Ecology 35, 610-621.

Engelking, B., Flessa, H., Joergensen, R.G., 2007. Shifts in amino sugar and ergosterol contents after addition of sucrose and cellulose to soil. Soil Biology & Biochemistry 39, 2111-2118.

Flessa, H., Potthoff, M., Loftfield, N., 2002. Laboratory estimates of CO_2 and N_2O emissions following surface application of grass mulch: importance of indigenous microflora of mulch. Soil Biology & Biochemistry 34, 875-879.

Frey, J.C., Pell, A.N., Berthiaume, R., Lapierre, H., Lee, S., Ha, J.K., Mendell, J.E., Angert, E.R., 2010. Comparative studies of microbial populations in the rumen, duodenum, ileum and faeces of lactating dairy cows. Journal of Applied Microbiology 108, 1982-1993.

Frostegård, Å, Petersen, S.O., Bååth, E., Nielsen, T.H., 1997. Dynamics of a microbial community associated with manure hot spots as revealed by phospholipid fatty acid analyses. Applied and Environmental Microbiology 63, 2224–2231.

Gattinger, A., Bausenwein, U., Bruns, C., 2004. Microbial biomass and activity in composts of different composition and age. Journal of Plant Nutrition and Soil Science 167, 556-561.

Gattinger, A., Höfle, M.G., Schloter, M., Embacher, A., Böhme, F., Munch, J.C., Labrenz, M., 2007. Traditional cattle manure application determines abundance, diversity and activity of methanogenic *Archaea* in arable European soil. Environmental Microbiology 9, 612–624.

Griffith, G.W., Ozkose, E., Theodorou, M.K., Davies, D.R., 2009. Diversity of anaerobic fungal populations in cattle revealed by selective enrichment culture using different carbon sources. Fungal Ecology 2, 87 –97.

Guerrero, C., Moral, R., Gómez, I., Zornoza, R., Arcenegui, V., 2007. Microbial biomass and activity of an agricultural soil amended with the solid phase of pig slurries. Bioresource Technology 98, 3259–3264.

Heinze, S., Raupp, J., Joergensen, R.G., 2010. Effects of fertilizer and spatial heterogeneity in soil pH on microbial biomass indices in a long-term field trial of organic agriculture. Plant and Soil 328, 203-215.

Jenkinson, D.S., 1988. The determination of microbial biomass carbon and nitrogen in soil. In: Wilson, J.R., (Ed.), Advances in Nitrogen Cycling in Agricultural Ecosystems. CABI, Wallingford, pp. 368-386.

Indorf, C., Dyckmans, J., Khan, K.S., Joergensen, R.G., 2011. Optimisation of amino sugar quantification by HPLC in soil and plant hydrolysates. Biology and Fertility of Soils 47, 387-396.

Joergensen, R.G., 1996. The fumigation-extraction method to estimate soil microbial biomass: calibration of the k_{EC} value. Soil Biology & Biochemistry 28, 25-31.

Joergensen, R.G., 2010. Organic matter and micro-organisms in tropical soils. In: Dion, P. (Ed.), Soil Biology and Agriculture in the Tropics. Springer, Berlin, pp. 17-44.

Joergensen, R.G., Emmerling, C., 2006. Methods for evaluating human impact on soil microorganisms based on their activity, biomass, and diversity in agricultural soils. Journal of Plant Nutrition and Soil Science 169, 295-309.

Joergensen, R.G., Mueller, T., 1996. The fumigation-extraction method to estimate soil microbial biomass: calibration of the k_{EN} value. Soil Biology & Biochemistry 28, 33-37.

Joergensen, R.G., Scheu, S., 1999. Depth gradients of microbial and chemical properties in moder soils under beech and spruce. Pedobiologia 43, 134-144.

Joergensen, R.G., Wichern, F., 2008. Quantitative assessment of the fungal contribution to microbial tissue in soil. Soil Biology & Biochemistry 40, 2977-2991.

Joergensen, R.G., Figge, R.M., Kupsch, L., 1997. Microbial decomposition of fuel oil after compost addition to soil. Zeitschrift für Pflanzenernährung und Bodenkunde 160, 21-24.

Joergensen, R.G., Wu, J., Brookes P.C., 2011. Measuring soil microbial biomass using an automated procedure. Soil Biology & Biochemistry 43, 873-876.

Kagami, M., von Elert, E., Ibelings, B.W., De Bruin, A., Van Donk, E., 2007. The parasitic chytrid, *Zygorhizidium*, facilitates the growth of the cladoceran zooplankter, *Daphnia*, in cultures of the inedible alga, *Asterionella*. Proceedings of the Royal Society B: Biological Sciences 274, 1561-1566.

Van Kessel, J.S., Pachepsky, Y.A., Shelton, D.R., Karns, J.S., 2007. Survival of *Escherichia coli* in cowpats in pasture and in laboratory conditions. Journal of Applied Microbiology 103, 1122–1127.

Klamer, M., Bååth, E., 2004. Estimation of conversion factors for fungal biomass determination in compost using ergosterol and PLFA 18:2ω6,9. Soil Biology & Biochemistry 36, 57-65.

Kyvsgaard, P., Sørensen, P., Møller, E., Magid, J., 2000. Nitrogen mineralization from sheep faeces can be predicted from the apparent digestibility of the feed. Nutrient Cycling in Agroecosystems 57, 207–214.

Leckie, S.E., Prescott, C.E., Grayston, S.J., Neufeld, J.D., Mohn, W.W., 2004. Comparison of chloroform fumigation-extraction, phospholipid fatty acid, and DNA methods to determine microbial biomass in forest humus. Soil Biology & Biochemistry 36, 529–532.

Lovell, R.D., Jarvis, S.C., 1996. Effect of cattle faeces on soil microbial biomass C and N in a permanent pasture soil. Soil Biology & Biochemistry 28, 291-299.

Ma, X., Wang, S., Jiang, G., Haneklaus, S., Schnug, E., Nyren, P., 2007. Short-term effect of targeted placements of sheep excrement on grassland in Inner Mongolia on soil and plant parameters. Communications in Soil Science and Plant Analysis 38, 1589-1604.

Mäder, P., Fließbach, A., Dubois, D., Gunst, L., Fried, P., Niggli, U., 2002. Soil fertility and biodiversity in organic farming. Science 296, 1694-1697.

Mueller, T., Joergensen, R.G., Meyer, B., 1992. Estimation of soil microbial biomass C in the presence of living roots by fumigation-extraction. Soil Biology & Biochemistry 24, 179-181.

Ouwerkerk, D., Klieve, A.V., 2001. Bacterial diversity within feedlot manure. Anaerobe 7, 59–66.

Pisz, J.M., Lawrence, J.R., Schafer, A.N., Siciliano, S.D., 2007. Differentiation of genes extracted from non-viable versus viable micro-organisms in environmental samples using ethidium monoazide bromide. Journal of Microbiological Methods 71, 312-318.

Rasul, G., Khan, K.S., Müller, T., Joergensen, R.G., 2008. Soil-microbial response to sugarcane filter cake and biogenic waste compost. Journal of Plant Nutrition and Soil Science 171, 355-360.

Rezaeian, M., Beakes, G.W., Parker, D.S., 2004a. Distribution and estimation of anaerobic zoosporic fungi along the digestive tracts of sheep. Mycological Research 34, 1227–1233.

Rezaeian, M., Beakes, G.W., Parker, D.S., 2004b. Methods for the isolation, culture and assessment of the status of anaerobic rumen chytrids in both in vitro and in vivo systems Mycological Research 34, 1215–1226.

Rezaeian, M., Beakes, G.W., Chaudhry, A.S., 2006. Effect of feeding chopped and pelleted lucerne on rumen fungal mass, fermentation profiles and *in sacco* degradation of barley straw in sheep. Animal Feed Science and Technology 128, 292–306.

Ritz, K., 2007. The plate debate: cultivable communities have no utility in contemporary environmental microbial ecology. FEMS Microbiology Ecology 60, 358–362.

Scheller, E., Joergensen, R.G., 2008. Decomposition of wheat straw differing in N content in soils under conventional and organic farming management. Journal of Plant Nutrition and Soil Science 171, 886-892.

Schnürer, J., 1993. Comparison of methods for estimating the biomass of three food-borne fungi with different growth patterns. Applied and Environmental Microbiology 59, 552-555.

Sekhavati, M.H., Mesgaran, M.D., Nassiri, M.R., Mohammadabadi, T., Rezaii, F., Fani Maleki, A., 2009. Development and use of quantitative competitive PCR assays for relative quantifying rumen anaerobic fungal populations in both in vitro and in vivo systems. Mycological Research 113, 1146-1153.

Siu, L., 2003. Mass of a bacterium. In: Elert, G. (Ed.), The Physics Factbook. http://hypertextbook.com/facts/2003/Louis.Siu.shtml

Sørensen, P, Jensen, E.S., 1995. Mineralization-immobilization and plant-uptake of nitrogen as influenced by the spatial distribution of cattle slurry in soils of different texture. Plant and Soil 173, 283–291.

Van Soest, P.J., 1967. Development of a comprehensive system of feed analyses and its application to forages. Journal of Animal Science 26, 119–128.

Strickland, M.S., Rousk, J., 2010. Considering fungal:bacterial dominance in soils – Methods, controls, and ecosystem implications. Soil Biology & Biochemistry 42, 1385-1395.

Taniwaki, M.H., Hocking, A.D., Pitt, J.I., Fleet, G.H., 2009. Growth and mycotoxin production by food spoilage fungi under high carbon dioxide and low oxygen atmospheres. International Journal of Food Microbiology 132, 100-108.

Vance, E.D., Brookes, P.C., Jenkinson, D.S., 1987. An extraction method for measuring soil microbial biomass C. Soil Biology & Biochemistry 19, 703-707.

Van Vliet, P.C.J., Reijs, J.W., Bloem, J., Dijkstra, J., de Goede, R.G.M., 2007. Effects of cow diet on the microbial community and organic matter and nitrogen content of feces. Journal of Dairy Science 90, 5146–5158.

Wachendorf, C., Joergensen, R.G., 2011. Mid-term tracing of ^{15}N derived from urine and dung in soil microbial biomass. Biology and Fertility of Soils 47, 147–155.

Weete, J.D., Weber, D.J., 1980. Lipid Biochemistry of Fungi and other Organisms. Plenum Publishing, New York.

Weete, J.D., Fuller, M.S., Huang, M.Q., Gandhi, S., 1989. Fatty acids and sterols of selected hyphochytriomycetes and chytridiomycetes. Experimental Mycology 13, 183-195.

Wolstrup, J., Jensen, K., 2008. Adenosine triphosphate and deoxyribonucleic acid in the alimentary tract of cattle fed different nitrogen sources. Journal of Applied Microbiology 45, 49–56.

Wu, J., Joergensen, R.G., Pommerening, B., Chaussod, R., Brookes, P.C., 1990. Measurement of soil microbial biomass C by fumigation-extraction - an automated procedure. Soil Biology & Biochemistry 22, 1167-1169.

Zelles, L., Hund, K., Stepper, K., 1987. Methoden zur relativen Quantifizierung der pilzlichen Biomasse im Boden. Zeitschrift für Pflanzenernährung und Bodenkunde 150, 249–252.

Zhao, X.R., Lin, Q., Brookes, P.C., 2005. Does soil ergosterol concentration provide a reliable estimate of soil fungal biomass? Soil Biology & Biochemistry 37, 311–317.

5. Microbial biomass in faeces of dairy cows affected by a nitrogen deficient diet

Daphne Isabel Jost[a]*, Martina Aschemann[b], Peter Lebzien[b], Rainer Georg Joergensen[c], Albert Sundrum[a]

[a] Department of Animal Nutrition and Animal Health, University of Kassel, Nordbahnhofstr. 1a, 37213 Witzenhausen, Germany

[b] Friedrich-Loeffler-Institute (FLI), Federal Research Institute for Animal Health, Institute of Animal Nutrition, Bundesallee 50, 38116 Braunschweig

[c] Department of Soil Biology and Plant Nutrition, University of Kassel, Nordbahnhofstr. 1a, 37213 Witzenhausen, Germany

Abstract

Faeces of cattle are an important nutrient source for plant growth but simultaneously the starting point of gaseous emissions and nutrient losses. Strategies to minimize emissions and nutrient losses into the environment require improved quantitative information on C and N fractions in faeces and their potential for faecal N mineralisation for plant uptake and emission, respectively. Since more than half of the faecal nitrogen originates from microbial N, the objective of the study was to develop a method for quantitatively detecting microbial biomass and portion of living microorganisms in dairy cattle faeces, including bacteria, fungi, and archaea. Three techniques were tested for repeatability. The chloroform fumigation extraction (CFE) method was applied for determining microbial biomass in cattle faeces. Detection of the fungal cell-membrane component ergosterol and amino sugar analysis of the cell-wall components fungal glucosamine and bacterial muramic acid served as indices for the microbial community structure. They also provided independent control values for the reliability of the microbial biomass range obtained by the CFE method. In a second step, an N deficient (ND) and an N balanced diet (NB) were compared with respect to the impacts on faecal C and N fractions, microbial indices and digestibility.

The CFE method in combination with fungal ergosterol and amino sugar analysis showed an acceptable reproducibility when applied in the analysis of faeces. The mean values of microbial biomass C and N concentrations averaged around 37 and 4.9 mg g^{-1} DM, respectively. Ergosterol, together with fungal glucosamine and bacterial muramic acid, revealed a percentage of 25% fungal

---

* Corresponding author. Tel.: +49 5542 98 1523; e-mail: daphne.jost@arcor.de

C in relation to the total microbial C content in dairy cattle faeces. Changes in ruminal N supply showed significant effects on faecal composition. Faecal concentrations of NDF, hemicelluloses and undigested dietary N (UDN) as well as the total C/N ratio were significantly higher in ND compared to the NB treatment. N deficiency was reflected also by a higher microbial biomass C/N ratio. It is concluded that the assessment of microbial indices provides valuable information with respect to diet effects on faecal composition and the successive decomposition. Further studies should be conducted to explore the potentials for minimising nutrient losses from faeces.

Keywords: cattle faeces, microbial biomass, bacterial and fungal distribution, N deficiency

5.1 Introduction

Conversion of ingested nutrients by ruminants into saleable products is a comparably inefficient process, as only an average of 25% of dietary protein is converted into meat and/or milk (Dewhurst et al., 2000; Calsamiglia et al., 2010). Nutrients in animal excreta that are not used by plants for growth processes contribute considerably to soil and ground water pollution (Petersen et al., 2007). Optimisation of the feeding regime has often been described as a key strategy for decreasing the portion of nutrients in the manure (Tamminga, 2003; Dijkstra et al., 2007). In the past, investigations of feeding effects on manure composition have mainly focused on slurry (Van Vliet et al., 2007), while little is published about the characteristics of solid manure. Gaseous emissions of N_2O and NH_3 are affected to a high degree by the availability of easily soluble faecal N and C fractions and by interactions with local environmental conditions (James et al., 1999; Arriaga et al., 2010). Organic matter with a high C/N ratio is expected to have long-term beneficial effects on soil biological functioning and nutrient use efficiency of plants (Haynes and Naidu, 1998; Mäder et al., 2002). For plant uptake, faecal N has to be mineralised after manure application. Mineralisation of faecal N is influenced, among other things, by variation in the fibre fractions (NDF, ADF, and ADL) and by the C/N ratio (Kyvsgaard et al., 2000). The organic bound N fraction is mineralised more rapidly in the soil than the fibre bound N (Sørensen et al., 2003; Powell et al., 2006).

Differences in the concentration of nutrients and organic components in the diet also affect the microbial community of the faeces (Van Vliet et al., 2007). As more than half of the faecal nitrogen originates from microbial N (Larsen et al., 2001), knowledge of the microbial biomass content in faeces is important to assess the efficiency of nutrient use. However, methods for quantifying microbial biomass and living microorganisms in fresh faeces are rare. With the exception of pathogens and coliform bacteria shed in faeces (Van Kessel et al., 2007), microbial populations of the intestinal tract and their relationship to host diet remain less well characterized (Frostegård et al., 1997; Gattinger et al., 2007; Van Vliet et al., 2007). The intestinal tract of cattle is colonized by a highly dynamic community of bacteria, archaea and fungi. Methods that enable the assessment of the total microbial biomass in faeces provide valuable information on the dynamics of fermentation processes and microbiological populations within the large intestine and after defecation.

In general, plate count approaches discriminate archaea and other non-cultivable microorganisms, which contribute more than 80% to the total number of species (Ouwerkerk and Klieve, 2001). Direct microscopic approaches often severely underestimate fungi (Joergensen and Wichern, 2008). New molecular techniques have made it possible to quantify certain microbial species or genera in different substrates and also in intestinal contents of cattle (Frey et al., 2010). The extraction of DNA followed by analysis of its composition provides important information on the microbial community composition of faeces (Van Vliet et al., 2007; Sekhavati et al., 2009), but usually reveals no information on the biomass, due to losses during extraction and unknown or highly variable DNA concentrations within different microorganisms (Leckie et al., 2004; Joergensen and Emmerling, 2006).

In contrast to the methods listed above, the chloroform fumigation extraction (CFE) method is able to estimate the biomass of all microorganisms surrounded by an intact cell-membrane, i.e. bacteria, fungi, and archaea. Additionally, the method has proven to facilitate the differentiation between living and dead microbial tissue in compost (Gattinger et al., 2004), pig manure (Aira et al., 2006; Guerrero et al., 2007) and cattle faeces (Jost et al., 2011). Ergosterol is an important component of fungal cell membranes, responsible for their stability and an index for fungal biomass (Weete and Weber, 1980; Djajakirana et al., 1996).

The aim of the present study was to gain quantitative information on faecal microbial biomass and the proportion of bacteria and fungi in faeces of dairy cows and to reflect changes in dietary N supply on faecal microbial composition.

5.2 Materials and Methods

5.2.1. Feeding regime

Faeces samples were taken from 10 dairy cows (German Holstein) that were part of a digestion trial on the experimental station of the Institute of Animal Nutrition of the Friedrich-Loeffler-Institute, Braunschweig, Germany (Aschemann et al., personal communication). The experiment comprised two feeding regimes with five cows each (n = 5). Faecal samples deriving from the same animal and regime were taken as representative for repeated measure (n = 3). All cows were fed with maize-based silage (*Zea mays* L.) and a concentrate consisting of 20% soybean (*Glycine max*

(L.) Merr.), 22.7% barley (*Hordeum vulgare* L.), 22.7% wheat (*Triticum aestivum* L.), 18.8% maize, 14.8% sugar beet (*Beta vulgaris* L.) pulp, and 1% mineral/vitamin mix. Diet composition and feed intake of silage and concentrate in kg d^{-1} are presented in Table 1. The experimental treatment comprised an N deficient diet (ND) and a control feeding with an optimal amount of rumen available N (NB), adjusted by addition of 3% urea to the concentrate.

The samples were taken rectally during three sampling days in May to July 2010, homogenized and immediately frozen in liquid nitrogen. This preservation technique has been tested, comparing the faeces samples obtained this way with fresh samples, which were immediately used for analysis and showed no significant differences (data not shown). A sub-sample was dried for 72 h at 60 °C and finely ground for chemical analyses.

Table 1. Feed intake of silage and concentrate in kg per day and diet composition

	Silage	Concentrate	
		ND	NB
Feeding (kg d^{-1})	9.9 ± 0.22	7.9 ± 0.18	7.9 ± 0.18
Dry matter (% FM)	36	88	88
Crude ash (g kg^{-1} DM)	39	44	42
Crude protein (g kg^{-1} DM)	74	190	275
Crude lipid (g kg^{-1} DM)	32	26	26
NDF (g kg^{-1} DM)	429	197	196
ADF (g kg^{-1} DM)	221	83	82
Urea (%)			3

ND = N deficient feeding, NB = N balanced feeding; DM = dry matter; FM = fresh matter; NDF = Neutral detergent fibre, ADF = Acid detergent fibre; cow replicates per feeding regime n = 5, replicate sub-samples for experiments n = 3; ± standard deviation.

5.2.2. Faecal C and N fractions

Total C and N were determined after dry combustion by gas chromatography using a Vario MAX (Elementar, Hanau, Germany) elemental analyzer. Total phosphor and sulphur were analyzed after HNO_3-pressure digestion as described by Chander et al. (2008) by ICP-AES (Spectro Analytical Instruments, Kleve, Germany). Faeces pH was measured using a 1/1 Aqua dist./faeces ratio. Crude ash was determined by ashing overnight at 550 °C in a muffle furnace. Ammonium and the organic components such as acid detergent lignin (ADL), acid detergent fibre (ADF), neutral detergent fibre (NDF), and undigested dietary N (UDN) were determined by near-infrared spectroscopy (FOSS 6500, Rellingen, Germany), as described by Althaus and Sundrum (University of Kassel, Witzenhausen, Germany, personal communication). Hemicellulose was calculated as the difference between NDF and ADF. All data were the mean of five cows for each feeding regime. Three replicates were taken from each cow on three different sampling days.

5.2.3. Microbial biomass C and N

Microbial biomass C and N were estimated by the chloroform fumigation extraction method (Brookes et al., 1985, Vance et al., 1987), modified by Jost et al. (2011). Two freshly thawed subsamples equivalent to 0.5 g oven-dry faeces were taken for the analysis. One sub-sample was fumigated at 25 °C with ethanol-free $CHCl_3$, which was removed after 24 h. Fumigated and non-fumigated portions were extracted with 100 ml 0.05 M $CuSO_4$ for 30 min by horizontal shaking at 200 rev min^{-1}. Following centrifugation (2000 g for 10 min), the faeces extract was filtered (filter paper hw3, Sartorius Stedim Biotech, Göttingen, Germany). Organic C in the extracts was measured as CO_2 by infrared absorption after combustion at 850 °C using a Dimatoc 100 automatic analyzer (Dimatec, Essen, Germany). Microbial biomass C was calculated as follows: microbial biomass C = E_C/k_{EC}, where E_C = (organic C extracted from fumigated faeces) − (organic C extracted from non-fumigated faeces) and k_{EC} = 0.45 (Wu et al., 1990; Joergensen, 1996). Total N in the extracts was measured by chemoluminescence detection. Microbial biomass N was calculated as follows: microbial biomass C = E_N/k_{EN}, where E_N = (organic N extracted from fumigated faeces) − (organic N extracted from non-fumigated faeces) and k_{EN} = 0.54 (Brookes et al., 1985; Joergensen and Mueller, 1996).

Together with the detection of microbial biomass by the CFE method, ergosterol was determined as an index for fungal biomass and as an independent control method. Determination of fungal glucosamine and bacterial muramic acid served as further independent control values for both CFE microbial biomass and ergosterol data (Appuhn and Joergensen, 2006; Jost et al., 2011).

5.2.4. Ergosterol analysis

The extraction method of Zelles et al. (1987) was used for determination of the fungal cell-membrane ergosterol. Freshly thawed faeces of 0.5 g DM were placed into 30 ml test tubes and treated with 10 ml methanol, 2.5 ml ethanol and 1 g KOH. The sample was saponified for 90 min at 70 °C under reflux. After cooling, ergosterol was extracted in two steps with 15 + 10 ml petroleum ether. From the supernatant, 15 ml were evaporated in a vacuum rotary evaporator at 40 °C. The non-polar fraction was dissolved in 5 ml methanol and stored at 4 °C until measurement. Ergosterol was determined by reversed-phase HPLC with 100% methanol as the mobile phase and detected at a wavelength of 282 nm (Jost et al., 2011).

5.2.5. Amino sugars

The amino sugars muramic acid, mannosamine, glucosamine and galactosamine were determined according to Indorf et al. (2011). Moist samples of 2 g fresh faeces were weighed into 20 ml test tubes, mixed with 10 ml 6 M HCl, and heated for 2 h at 105 °C. After HCl removal from the filtered hydrolysates in a vacuum rotary evaporator at 40 °C and centrifugation, the samples were transferred to vials and stored at -18 °C until the HPLC measurements. Chromatographic separations were performed on a Phenomenex (Aschaffenburg, Germany) Hyperclone C_{18} column (125 mm length × 4 mm diameter), protected by a Phenomenex C_{18} security guard cartridge (4 mm length × 2 mm diameter) at 35 °C. The HPLC system consisted of a Dionex (Germering, Germany) P 580 gradient pump, a Dionex Ultimate WPS – 3000TSL analytical autosampler, with in-line split-loop injection and thermostat, and a Dionex RF 2000 fluorescence detector set at 445 nm emission and 330 nm excitation wavelengths, with medium sensitivity. OPA reagent (*o*-Phthalaldehyde), sample and preparation vials were stored in the autosampler at 15 °C. For the automated pre-column derivatisation, 50 µl OPA and 30 µl sample were mixed in the preparation vial, and 15 µl of the indole derivates were injected after 120 sec reaction time. The mobile phase consisted of two eluents and was delivered at a flow rate of 1.5 ml min^{-1}. Eluent A was a 97.8/0.7/1.5 (v/v/v) mixture of an aqueous phase, methanol and tetrahydrofuran (THF). The aqueous phase contained 52 mmol sodium citrate and 4 mmol sodium acetate, adjusted to pH 5.3 with HCl. Methanol and THF were then added. Eluent B consisted of 50% water and 50% methanol (v/v).

Fungal C (mg g^{-1} dry weight) was calculated by subtracting bacterial glucosamine from total glucosamine as an index for fungal residues, assuming that muramic acid and glucosamine occur at a 1 to 2 molar ratio in bacterial cells (Engelking et al., 2007): mg fungal C g^{-1} dry weight = (mmol glucosamine − mmol muramic acid) × 179.2 g mol^{-1} × 9, where 179.2 is the molecular weight of glucosamine and 9 the conversion value of fungal glucosamine to fungal C (Appuhn and Joergensen, 2006). Bacterial C (µg g^{-1} dry weight) was calculated as an index for bacterial residues by multiplying the concentration of muramic acid in µg g^{-1} dry weight by 45 (Appuhn and Joergensen, 2006).

5.2.6. Statistical analysis

The results presented in the tables are arithmetic means and expressed on an oven-dry basis (about 72 h at 60°C). The significance of difference was tested by one-way analysis of variance for repeated measures. All statistical analyses were performed using JMP 7.0 (SAS Inst. Inc.).

5.3 Results

The dry matter percentage of faeces samples ranged between 13 and 17 % (data not shown). Mean organic C (data not shown) and N content averaged 604 and 29 mg g^{-1} DM, respectively, without significant differences between treatments (Table 2). The mean faecal pH-value was significantly higher in NB (6.7, $P<0.01$) compared to ND (6.2), and was accompanied by a significantly lower C/N ratio of 20 ($P = 0.04$) in this treatment compared with ND (23). Concentrations of hemicellulose, neutral detergent fibre and undigested dietary N were significantly higher, while the concentration of crude ash was significantly lower in the faeces of ND compared with NB ($P<0.01$). NH_4^+ and the organic components ADF, ADL did not differ significantly between the treatments. Significant day × feeding interactions were observed for pH, total N, NH_4^+, and hemicellulose.

Table 2. Elemental composition and organic components in the faeces of dairy cows fed an N deficient and an N balanced diet

	pH (H$_2$O)	N	C/N	NH_4^+	Crude ash	Hemi-cellulose	NDF	ADF	ADL	UDN
		(mg g^{-1} DM)				(mg g^{-1} DM)				
Mean	6.4	29	22	0.68	75	322	668	346	58	0.78
ND	6.2	27	23	0.61	64	370	710	337	52	0.83
NB	6.7	30	20	0.74	86	270	630	355	64	0.73
Probability values										
Feeding	0.01	ns	0.04	ns	<0.01	<0.01	0.01	ns	ns	0.01
Day	ns	ns	ns	<0.01	ns	ns	ns	ns	ns	ns
Day × feeding	0.02	0.04	ns	0.04	ns	0.03	ns	ns	ns	ns
CV (±%)	23	26	15	41	21	20	11	13	48	13

DM = dry matter, ns = non-significant; ND = N deficient feeding, NB = N balanced feeding; CV = pooled coefficient of variation between feeding regimes (cow replicates n = 5, replicate sub-samples n = 3); significant differences tested with ANOVA for repeated measures; NDF = neutral detergent fibre, ADF = acid detergent fibre; ADL = acid detergent lignin; UDN = undigested dietary N

With the CFE method, a mean faecal content of C and N of 37 and 4.9 mg g^{-1} DM, respectively, was detected in the microbial biomass without showing a significant day to day variation (Table 3). Concentrations of microbial biomass N in the cattle faeces varied around 4.0 mg g^{-1} DM for ND and 5.8 mg g^{-1} DM for NB, without being affected significantly by the treatment. The mean microbial biomass C/N ratio reflected N deficiency in the diet with 9.1 and 7.0 for ND and NB, respectively ($P<0.01$, Table 4). The microbial biomass C content in faeces of dairy cows contributed more than 6% to the total C, while over 17% of the total faecal N were derived from microbial biomass N.

Table 3. Microbial biomass, amino sugar indices and microbial C in faeces of dairy cows fed an N deficient and an N balanced diet

	Microbial biomass		Ergosterol	MurN	ManN	GlcN	GalN	Microbial C
	C	N						
	(mg g^{-1} DM)		(µg g^{-1} DM)	(mg g^{-1} DM)				
Mean	37	4.9	13.2	0.80	0.22	2.4	2.1	50
ND	35	4.0	12.7	0.76	0.19	2.4	2.1	50
NB	39	5.8	13.7	0.84	0.24	2.4	2.0	49
Day 1	37	4.5	9.3	0.60	0.24	1.9	1.7	36
Day 2	35	5.0	15.2	0.71	0.17	2.1	1.9	42
Day 3	39	5.2	15.2	1.11	0.30	3.2	2.7	65
Probability values								
Feeding	ns	ns	ns	ns	ns	ns	ns	ns
Day	ns	ns	0.01	0.01	<0.004	0.01	0.01	0.01
Day × feeding	ns	ns	ns	ns	ns	ns	ns	ns
CV (±%)	34	35	36	39	36	35	35	37

DM = dry matter; ND = N deficient feeding, NB = N balanced feeding; ns = non-significant; MurN = muramic acid; ManN = mannosamine; GlcN = glucosamine; GalN = galactosamine; CV = pooled coefficient of variation between feeding regimes (cow replicates n = 5, replicate sub-samples n = 3); significant differences tested with a one-way ANOVA for repeated measures

Table 4. Microbial ratios in faeces of dairy cows fed an N deficient and an N balanced diet

	Microbial biomass C/N	Ergosterol/ microbial biomass C (‰)	Fungal C/ bacterial C	Fungal glucosamine/ ergosterol
Mean	8.0	0.35	0.34	107
ND	9.1	0.36	0.36	119
NB	7.0	0.34	0.31	94
Probability values				
Feeding	<0.01	ns	ns	0.03
Day	ns	ns	ns	ns
Day × feeding	ns	ns	<0.01	ns
CV (±%)	26	40	30	32

ND = N deficient feeding, NB = N balanced feeding; ns = non-significant; CV = pooled coefficient of variation between feeding regimes (cow replicates n = 5, replicate sub-samples n = 3); significant differences tested with a one-way ANOVA for repeated measures

The mean concentrations of muramic acid, mannosamine, glucosamine, and galactosamine amounted to 0.80, 0.22, 2.4 and 2.1 mg g^{-1} DM, respectively. Mean ergosterol content reached 13 µg g^{-1} DM. The resulting ergosterol to microbial biomass C ratio resulted in 0.35‰. On the basis of the amino sugar data, a mean microbial C concentration of 49 mg g^{-1} DM was calculated from bacterial muramic acid and fungal glucosamine, with a mean percentage of 25% fungal C and 75% bacterial C. Faecal microbial biomass C determined by the CFE method corresponded to 75% of the total microbial C calculated from the amino sugar data. Except for NH_4^+, ergosterol and amino sugars, no significant day to day variation was found.

Microbial biomass N correlated significantly with the crude protein content of the faeces ($r = 0.55$, Fig. 1). Fungal C and ergosterol showed a significant positive linear relationship ($r = 0.43$, Fig. 2), whereas muramic acid was negatively correlated with the total N content ($r = -0.57$, Fig. 3). Significant day x feeding interactions ($P<0.01$) were found for the fungal to bacterial C ratio. The fungal glucosamine to ergosterol ratio was significantly lower for NB than for ND, at 57 vs. 76 ($P = 0.03$).

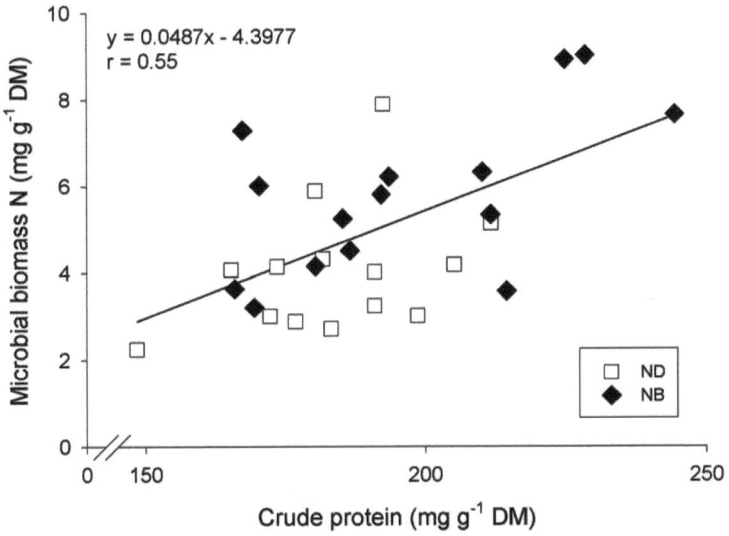

Fig. 1. Linear relationship between crude protein and microbial biomass N in faeces of dairy cows in relation to the different treatments ($r = 0.55$, n = 30, $P < 0.01$).

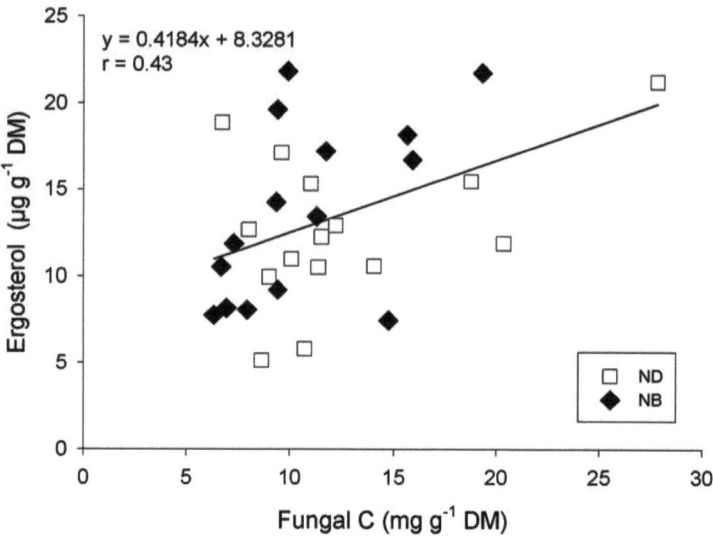

Fig. 2. Linear relationship between fungal C and ergosterol in faeces of dairy cows in relation to the different treatments ($r = 0.43$, n = 30, $P < 0.01$).

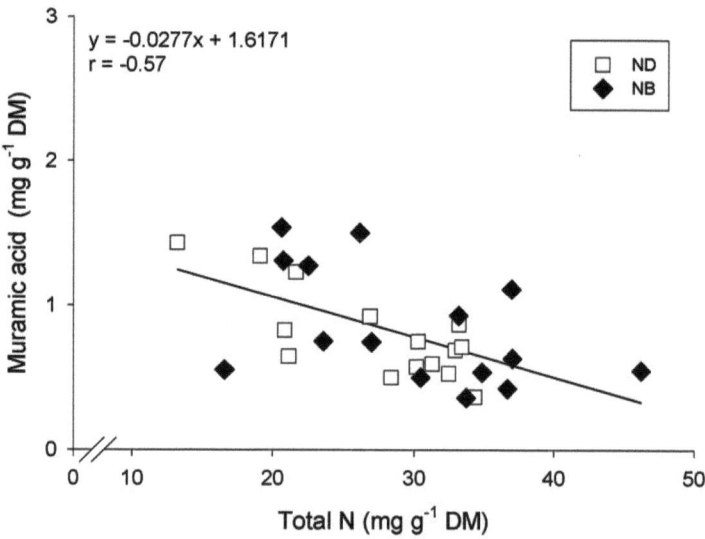

Fig. 3. Linear relationship between bacterial muramic acid and total N in faeces of dairy cows in relation to the different treatments ($r = -0.57$, n = 30, $P < 0.01$).

5.4 Discussion

5.4.1. Microbial indices

The indicators of microbial biomass showed a low intra-individual variation between sampling days, indicating a good reproducibility of the data. A low variability of the parameters within the same treatment has been obtained previously by Jost et al. (2011) in faeces of heifers. However, the values found in heifers differed considerably from the values assessed in lactating dairy cows, although analyzed by the same set of methods. C and N content of microbial biomass determined in the faeces from heifers were three times lower than in the faeces of dairy cows in the current study, at 9.3 and 1.9 mg g^{-1} DM, respectively. Additionally, mean microbial biomass C/N ratio showed lower values in faeces of heifers compared with lactating dairy cows. These differences are most likely due to differences in diet composition, with dairy cows receiving high amounts of concentrate, providing a higher substrate source for intestinal microbial growth and affecting faecal bacterial biomass (Van Vliet et al., 2007).

In the current study, the concentration of microbial C, calculated on the basis of bacterial muramic acid and fungal glucosamine was approximately 50% higher compared with those in heifers (Jost et al., 2011). According to Amelung (2001, 2008), the cell-wall components muramic acid and glucosamine have the tendency to be accumulated in microbial residues, suggesting that roughly 75% of the faecal microbial C in the present study belong to the living fraction and the other 25% are within the remains of dead fungi and bacteria. In the highly dynamic situation of C and N supply in the gut, rapid microbial growth is probably accompanied by concomitant microbial death, as observed in soil (Chander and Joergensen, 2001). Although the current figures seem to be realistic, they should be compared with other methods, because it is not known (1) whether the glucosamine and muramic acid concentrations are identical in living and dead microbial tissue and (2) whether the several conversion values for microbial biomass C by the fumigation extraction method and for microbial C by amino sugar analysis may lead to an underestimation of microbial biomass C. Furthermore, this approach neglects the significant contribution of archaea to the faecal microbial biomass, which contribute approximately 16% of the archaeal phospholipid etherlipid (PLEL) to the total phospholipid chain content in cattle manure (Gattinger et al., 2007). Archaea do not contain the amino sugar muramic acid (Kandler and König, 1998), but most likely add to the CHCl$_3$ labile C fraction obtained by the fumigation extraction method.

In the context of the urea application in the N balanced diet, a decline was observed in the microbial biomass C/N ratio and the fungal C to ergosterol ratio. The higher supply of easily available N is expected to increase the N storage in faecal microbial biomass, apparently accompanied by a shift within the fungal community. Potential differences in microbial biomass content between the treatments were most likely minimized due to higher N use efficiency with lower N availability from the diet (Bach et al., 2005; Calsamiglia et al., 2010).

5.4.2. Bacterial and fungal contribution to microbial tissue

The average fungal C to bacterial C ratio of 0.34 found in this study is in line with the assumption of 25% fungal C and 75% bacterial C. This means that the faeces of dairy cows contained considerably less fungal C than faeces of heifers, where fungi contributed more than 40% to microbial C (Jost et al., 2011). This difference is due to the higher content of structural components such as ADF in the heifer diet.

Chitin, the polymer of glucosamine, has previously been used as an indicator of fungal biomass in rumen fluid of cattle (Sekhavati et al., 2009) and in sheep faeces (Rezaeian et al., 2004a, b). The chitin concentration in sheep faeces was 10.2 mg g^{-1} DM (Rezaeian et al., 2004a). In a sheep feeding experiment, the liquid associated fraction of rumen material contained 12.1 to 15.4 mg chitin g^{-1} DM and in the corresponding particle associated fraction 2.6 to 3.9 mg chitin g^{-1} DM. One reason for the markedly higher chitin concentration compared with the glucosamine concentration in the present study might be the differences in rumen and intestine physiology between sheep and cattle. However, Rezaeian et al. (2004a, 2006) used the colorimetric assay of Chen and Johnson (1983), which cannot distinguish between fungal glucosamine, bacterial glucosamine, and galactosamine. This may lead to an overestimation of fungal tissue. In accordance with the present data, Rezaeian et al. (2004a) have estimated that anaerobic fungi comprise approximately 20% of the total microbial biomass in the rumen of sheep.

In the current study, the mean faecal concentration of ergosterol was 13.2 µg g^{-1} DM. Klamer and Bååth (2004) obtained a factor of 190 when calculating the fungal biomass C from the ergosterol concentration in 11 compost fungi species. Taking this factor into account would result in a mean fungal biomass C content of 2.5 mg C g^{-1} DM in the present faeces samples, suggesting that ergosterol containing fungi contribute a proportion of 7% to the total microbial biomass in both treatments.

Ergosterol has been repeatedly used as an index for fungal biomass in soil (Bååth and Anderson, 2003) and other solid substrates (Newell, 1992), but only once in cattle faeces (Jost et al., 2011). Of the anaerobic fungal populations, yeasts such as *Candida sp.* (Ahmad et al., 2010) or *Saccharomyces cerevisia* (Aguilera et al., 2006) contain high concentrations of ergosterol, also food spoiling *Mucor plumbeus* (Taniwaki et al., 2009). In contrast, no ergosterol but high concentrations of cholesterol were measured in chytridiomycetes (Weete et al., 1989; Kagami et al., 2007). Until now, no information has been available on the ergosterol concentration of anaerobic fungal species found in cattle, such as *Anaeromyces, Orpinomyces, Caecomyces*, or *Piromyces* (Griffith et al., 2009).

The majority of microbial data on faeces samples are based on colony forming units (CFU) or thallus forming units (TFU), counted on plates or as the most probable number (Griffith et al., 2009). Edwards et al. (2008) recovered just 11% of the bacterial taxa present within the rumen in culture. The bacterial biomass ranged from 1.2 to 8.0 mg C g^{-1} DM, depending on the quality of the diet (Van Vliet et al., 2007). These values are similar to those of Jost et al. (2011), but markedly below the present data. Nevertheless, it should be considered that fungal biomass was not included in this technique, using confocal laser scanning microscopy and europium chelate/fluorescent brightener differential stain in combination with automated image analysis (Bloem et al., 1994). According to Frostegård et al. (1997), cattle manure contains about 2.6 mol% linoleic acid (18:2ω6.9). If linoleic acid is converted to fungal biomass by multiplying by 107 (Joergensen and Wichern, 2008), total PLFA contents of 300 nmol (Frostegård et al., 1997) and 1500 nmol (Gattinger et al., 2007) corresponded to a fungal biomass C ranging from 0.8 to 4.2 mg g^{-1} DM. The results for fungal biomass C, based on ergosterol and glucosamine analysis, were within the ranges determined in the present study.

5.4.3. Digestibility affected by N deficient diet

N deficiency in the diet showed a significant effect on nutrient digestibility. Concentration of neutral detergent fibre, hemicelluloses, and undigested dietary N in faeces were significantly higher in ND compared with the NB treatment ($P<0.05$). The results are in line with those found by Ruiz et al. (2002), who investigated the effects of urea supplementation on nitrogen deficient diets. However, the results are in contrast to those of Cameron et al. (1991) and Boucher et al. (2007). In both investigations, the authors conclude that the supplementation of urea did not affect the digestibility of DM and NDF. The different conclusions might be due to the fact that, in contrast to the current study, no nitrogen deficient diets were used. Animal responses to dietary NPN supplements require that the available N in the rumen is below the microbial requirement (Van Soest, 1994). When ruminal N is deficient, fibre fermentation and microbial yield can be depressed. The supplementation of urea to the N-deficient diet is expected to increase the growth of rumen microbiota and their fermentation activity (Erdman et al., 1986; Griswold et al., 2003), which might have caused a better fermentation of NDF and hemicelluloses in the rumen and correspondingly a lower concentration in the faeces. The significant increase of crude ash concentration in the faeces of the NB treatment is interpreted as a result of the lower concentration of organic matter in the faeces.

C/N ratio in faeces of the NB treatment was significantly lower than in the ND treatment, which is primarily due to the differences in N availability in the rumen and the subsequent differences in microbial biomass in the hindgut. Hristov et al. (2004) did not find an effect of an increase in CP content in the diet on C/N ratio in faeces when adding protein to the diet of low producing dairy cows, already supplied with a level of nitrogen corresponding to microbial requirements. They explained the lack of animal responses with the lack of available ruminally degradable carbohydrates and the loss of N by urinary excretion.

The mean faecal pH values found in the NB (6.7) and in the ND treatment (6.2) were lower than those found in the studies of Ireland-Perry and Stalling (1993). While they made use of corn silage and orchard grass hay as roughage, only corn silage was used as roughage in the current study. In the NB treatment, mean faecal pH value (6.7) was significantly higher than in the ND treatment (6.2). The reason might be due to the differences in the concentration of hemicelluloses in the hindgut, which in the case of the ND treatment provided a higher substrate source for microbial fermentation and production of volatile fatty acids compared with the NB treatment. The buffering capacity in the hindgut is lower than that in the rumen (Gressley, 2011).

5.5 Conclusions

The selected methods provided insights into the variability of faecal composition, encompassing all living microorganisms in faeces. Determination of microbial biomass with the CFE method in combination with fungal ergosterol and amino sugar analysis as supplemental microbial indicators showed a good reproducibility of the methods. Consistent data confirm that the applied techniques are expected to be suitable tools for analysing the effects of diet composition on faeces or on other intestinal contents and the subsequent behaviour of microorganisms in soil. Especially ergosterol and amino sugar analysis should be used additionally in the rumen and intestines as indices for fungal and bacterial contribution to ruminant digestion processes. Dietary N deficiency is reflected by a higher microbial biomass C/N ratio and lower digestibility of NDF and hemicelluloses. It is concluded that the assessment of microbial indices provides valuable information with respect to effects of diets on faecal composition and the successive decomposition. Further studies including a larger number of cows and diet variants should be conducted with the aim of minimising nutrient losses and, thus, achieving better nutrient use efficiency.

Acknowledgments

The skilful technical assistance of Gabriele Dormann and Christiane Jatsch is highly appreciated. This project was supported by a grant of the Research Training Group 1397 "Regulation of soil organic matter and nutrient turnover in organic agriculture" of the German Research Foundation (DFG). The animal trial was carried out with financial support from EU FP7 REDNEX. Faeces samples were provided by staff members of the Friedrich-Loeffler-Institute (FLI), Federal Research Institute for Animal Health, Institute of Animal Nutrition (ITE), Braunschweig, for which we are very grateful.

5.6 References

Aguilera, F., Peinado, R.A., Millán, C., Ortega, J.M., Mauricio, J.C., 2006. Relationship between ethanol tolerance, H^+-ATPase activity and the lipid composition of the plasma membrane in different wine yeast strains. Int. J. Food Microbiol. 110, 34–42.

Ahmad, A., Khan, A., Manzoor, N., Khan, L.A., 2010. Evolution of ergosterol biosynthesis inhibitors as fungicidal against Candida. Microb. Pathog. 48, 35–41.

Aira, M., Monroy, F., Domínguez, J., 2006. Changes in microbial biomass and microbial activity of pig slurry after the transit through the gut of the earthworm Eudrilus eugeniae. Biol. Fertil. Soils 42, 371–376.

Amelung, W., 2001. Methods using amino sugars as markers for microbial residues in soil. In: Lal, J.M., Follett, R.F., Stewart, B.A. (Eds.), Assessment Methods for Soil Carbon. Lewis Publishers, Boca Raton, pp. 233–272.

Amelung, W., Brodowski, S., Sandhage-Hofmann, A., Bol, R., 2008. Combining biomarker with stable isotope analyses for assessing the transformation and turnover of soil organic matter. Advance in Agronomy 100, 155–250.

Appuhn, A., Joergensen, R.G., 2006. Microbial colonisation of roots as a function of plant species. Soil Biol. Biochem. 38, 1040–1051.

Arriaga, H., Salcedo, G., Calsamiglia, S., Merino, P., 2010. Effect of diet manipulation in dairy cow N balance and nitrogen oxides emissions from grasslands in northern Spain. Agric. Ecosyst. Environ. 135, 132–139.

Bååth, E., Anderson, T.H., 2003. Comparison of soil fungal/bacterial ratios in a pH gradient using physiological and PLFA-based techniques. Soil Biol. Biochem. 35, 955–963.

Bach, A., Calsamiglia, S., Stern, M.D., 2005. Nitrogen metabolism in the rumen. J. Dairy Sci. 88, E9-E21.

Bloem, J., Lebbink, G., Zwart, K.B., Bouwman, L.A., Burgers, S.L., De Vos, J.A., Ruiter, P.C., 1994. Dynamics of microorganisms, microbivores and nitrogen mineralisation in winter wheat fields under conventional and integrated management. Agric. Ecosyst. Environ. 51, 129–143.

Boucher, S.E., Ordway, R.S., Whitehouse, N.L., Lundy, F.P., Kononoff, P.J., Schwab, G.C., 2007. Effect of incremental urea supplementation of a conventional corn silage based diet on ruminal ammonia concentration and synthesis of microbial protein. J. Dairy Sci. 90, 5619–5633.

Brookes, P., Landman, A., Pruden, G., Jenkinson, D.S., 1985. Chloroform fumigation and the release of soil nitrogen: A rapid direct extraction method to measure microbial biomass nitrogen in soil. Soil Biol. Biochem. 17, 837–842.

Calsamiglia, S., Ferret, A., Reynolds, C.K., Kristensen, N.B., Van Vuuren, A.M., 2010. Strategies for optimizing nitrogen use by ruminants. Animal 4, 1184–1196.

Cameron, M.R., Klusmeyer, T.H., Lynch, G.L., Clark, J.H., 1991. Effects of urea and starch on rumen fermentation, nutrient passage to the duodenum, and performance of cows. J. Dairy Sci. 74, 1321–1336.

Chander, K., Hartmann, G., Joergensen, R.G., Khan, K.S., Lamersdorf, N., 2008. Comparison of methods for measuring heavy metals in soils contaminated by different sources. Archives of Agronomy and Soil Science 54, 413–422.

Chander, K., Joergensen, R.G., 2001. Decomposition of 14C glucose in two soils with different levels of heavy metal contamination. Soil Biol. Biochem. 33, 1811–1816.

Chen, G.C., Johnson, B.R., 1983. Improved colorimetric determination of cell wall chitin in wood decay fungi. Appl. Environ. Microbiol. 46, 13–16.

Dewhurst, R.J., Davies, D.R., Merry, R.J., 2000. Microbial protein supply from the rumen. Anim. Feed Sci. Technol. 85, 1–21.

Dijkstra, J., Kebreab, E., Mills, J.A., Pellikaaan, W.F., Lopez, S., Bannink, A., France, J., 2007. Predicting the profile of nutrients available for absorption: from nutrient requirement to animal response and environmental impact. Animal 1, 99–111.

Djajakirana, G., Joergensen, R., Meyer, B., 1996. Ergosterol and microbial biomass relationship in soil. Biol. Fertil. Soils 22, 299–304.

Edwards, J.E., Huws, S.A., Kim, E.J., Lee, M.R., Kingston-Smith, A.H., Scollan, N.D., 2008. Advances in microbial ecosystem concepts and their consequences for ruminant agriculture. Animal 2, 653–660.

Engelking, B., Flessa, H., Joergensen, R.G., 2007. Shifts in amino sugar and ergosterol contents after addition of sucrose and cellulose to soil. Soil Biol. Biochem. 39, 2111–2118.

Erdman, R.A., Proctor, G.H., Vanadersall, J.H., 1986. Effect of rumen ammonia concentration on in situ rate and extent of digestion of feedstuffs. J. Dairy Sci. 69, 2312–2320.

Frey, J.C., Pell, A.N., Berthiaume, R., Lapierre, H., Lee, S., Ha, J.K., Mendell, J.E., Angert, E.R., 2010. Comparative studies of microbial populations in the rumen, duodenum, ileum and faeces of lactating dairy cows. J. Appl. Microbiol. 108, 1982–1993.

Frostegård, A., Petersen, S.O., Bååth, E., Nielsen, T.H., 1997. Dynamics of a microbial community associated with manure hot spots as revealed by phospholipid fatty acid analyses. Appl. Environ. Microbiol. 63, 2224–2231.

Gattinger, A., Bausenwein, U., Bruns, C., 2004. Microbial biomass and activity in composts of different composition and age. J Plant Nutr Soil Sci 167, 556–561.

Gattinger, A., Höfle, M.G., Schloter, M., Embacher, A., Böhme, F., Munch, J.C., Labrenz, M., 2007. Traditional cattle manure application determines abundance, diversity and activity of methanogenic Archaea in arable European soil. Environ. Microbiol. 9, 612–624.

Gressley, T.F., Hall, M.B., Armentano, L.E., 2011. Ruminant nutrition symposium: productivity, digestion, and health responses to hindgut acidosis in ruminant. Animal Sci. 1989, 1120–1130.

Griffith, G.W., Ozkose, E., Theodorou, M.K., Davies, D.R., 2009. Diversity of anaerobic fungal populations in cattle revealed by selective enrichment culture using different carbon sources. Fungal Ecol 2, 87–97.

Griswold, K.E., Apgar, G.A., Bouton, J., Firkins, J.L., 2003. Effects of urea infusion and ruminal degradable protein concentration on microbial growth, digestibility, and fermentation in continuous culture. Animal Sci. 81, 329–336.

Guerrero, C., Morala, R., Gómeza, I., Zornozaa, R., Arceneguia, V., 2007. Microbial biomass and activity of an agricultural soil amended with the solid phase of pig slurries. Bioresour Technol 98, 3259–3264.

Haynes, R.J., Naidu, R., 1998. Influence of lime, fertilizer and manure application pm soil organic matter content and soil physical conditions: a review. Nutr. Cycl. Agroecosyst. 51, 123–137.

Hristov, A.N., Etter, R.P., Ropp, J.K., Grandeen, K.L., 2004. Effect of dietary crude protein level and degradability on ruminal fermentation and nitrogen utilization in lactating dairy cows. J. Anim. Sci. 82, 3219–3229.

Indorf, C., Dyckmans, J., Khan, K.S., Joergensen, R.G., 2011. Optimisation of amino sugar quantification by HPLC in soil and plant hydrolysates. Biol. Fertil. Soils.

Ireland-Perry, R.L., Stallings, C.C., 1993. Fecal consistency as related to dietary composition in lactating Holstein cows. J. Dairy Sci. 76, 1074–1082.

James, T., Mexer, D., Esparza, E., Depeters, E.J., Perez-Monti, H., 1999. Effects of dietary nitrogen manipulation on ammonia volatilization from manure from Holstein heifers. J. Dairy Sci. 82, 2430–2439.

Joergensen, R.G., 1996. The fumigation-extraction method to estimate soil microbial biomass: Calibration of the kEC value. Soil Biol. Biochem. 28, 25–31.

Joergensen, R.G., Emmerling, C., 2006. Methods for evaluating human impact on soil microorganisms based on their activity, biomass, and diversity in agricultural soils. J Plant Nutr Soil Sci 169, 295–309.

Joergensen, R.G., Mueller, T., 1996. The fumigation-extraction method to estimate soil microbial biomass: Calibration of the kEN value. Soil Biol. Biochem. 28, 33–37.

Joergensen, R.G., Wichern, F., 2008. Quantitative assessment of the fungal contribution to microbial tissue in soil. Soil Biol. Biochem. 40, 2977–2991.

Jost, D.I., Indorf, C., Joergensen, R.G., Sundrum, A., 2011. Determination of microbial biomass and fungal and bacterial distribution in cattle faeces. Soil Biol. Biochem. doi:10.1016/j.soilbio.2011.02.013,

Kagami, M., von Elert, E., Ibelings, B.W., De Bruin, A., Van Donk, E., 2007. The parasitic chytrid, Zygorhizidium, facilitates the growth of the cladoceran zooplankter, Daphnia, in cultures of the inedible alga, Asterionella. Proc. Biol. Sci. 274, 1561–1566.

Kandler, O., König, H., 1998. Cell wall polymers in Archaea (Archaebacteria). Cell. Mol. Life Sci. 54, 305–308.

Van Kessel, J.S., Pachepsky, Y.A., Shelton, D.R., Karns, J.S., 2007. Survival of Escherichia coli in cowpats in pasture and in laboratory conditions. J. Appl. Microbiol. 103, 1122–1127.

Klamer, M., Bååth, E., 2004. Estimation of conversion factors for fungal biomass determination in compost using ergosterol and PLFA 18:2ω6,9. Soil Biol. Biochem. 36, 57–65.

Kyvsgaard, P., Sørensen, P., Møller, E., Magid, J., 2000. Nitrogen mineralization from sheep faeces can be predicted from the apparent digestibility of the feed. Nutr. Cycl. Agroecosyst. 57, 207–214.

Larsen, M., Madsen, T.G., Weisbjerg, M.R., Hvelplund, T., Madsen, J., 2001. Small intestinal digestibility of microbial and endogenous amino acids in dairy cows. J. Anim. Physiol. Anim. Nutr. 85, 9–21.

Leckie, S.E., Prescott, C.E., Graystonb, S.J., Neufeldc, J.D., Mohnc, W.W., 2004. Comparison of chloroform fumigation-extraction, phospholipid fatty acid, and DNA methods to determine microbial biomass in forest humus. Soil Biol. Biochem. 36, 529–532.

Mäder, P., Fließbach, A., Dubois, D., Gunst, L., Fried, P., Niggli, U., 2002. Soil Fertility and Biodiversity in Organic Farming. Science 296, 1694–1697.

Newell, S.Y., 1992. Estimating fungal biomass and productivity in decomposing litter. In: G.C. Carol and D.T. Wicklow (Ed.), The Fungal Community: its Organization and Role in the Ecosystem, 2nd ed. Marcel Dekker, New York, pp. 521–561.

Ouwerkerk, D., Klieve, A.V., 2001. Bacterial Diversity within Feedlot Manure. Anaerobe 7, 59–66.

Petersen, S.O., Sommer, S.G., Béline, F., Burton, C., Dach, J., Dourmad, J.Y., Leip, A., 2007. Recycling of livestock manure in a whole-farm perspective. Livest. Sci. 112, 180–191.

Powell, J.M., Broderick, G.A., Grabber, J.H., Hymnes-Fecht, U.C., 2009. Technical note: Effects of forage protein-binding polyphenols on chemistry of dairy excreta. J. Dairy Sci. 92, 1765–1769.

Rezaeian, M., Beakes, G.W., Chaudhry, A.S., 2006. Effect of feeding chopped and pelleted lucerne on rumen fungal mass, fermentation profiles and in sacco degradation of barley straw in sheep. Anim. Feed Sci. Technol. 128, 292–306.

Rezaeian, M., Beakes, G.W., Parker, D.S., 2004b. Methods for the isolation, culture and assessment of the status of anaerobic rumen chytrids in both in vitro and in vivo systems. Mycol. Res. 108, 1215–1226.

Rezaeian, M., Beakes, G.W., Parker, D.S., 2004a. Distribution and estimation of anaerobic zoosporic fungi along the digestive tracts of sheep. Mycol. Res. 108, 1227–1233.

Ruiz, R., Tedeschi, L.O., Marini, J.C., Fox, D.G., Pell, A.N., Jarvis, G., Russell, J.B., 2002. The effect of ruminal nitrogen (N) deficiency in dairy cows: Evaluation of the Cornell Net Carbohydrate and Protein System ruminal N deficiency adjustment. J. Dairy Sci. 85, 2986–2999.

Sekhavati, M.H., Mesgaran, M.D., Nassiri, M.R., Mohammadabadi, T., Rezaii, F., Maleki, A.F., 2009. Development and use of quantitative competitive PCR assays for relative quantifying rumen anaerobic fungal populations in both in vitro and in vivo systems. Mycol. Res. 113, 1146–1153.

Sørensen, P., Weisbjerg, M.R., Lund, P., 2003. Dietary effects on the composition and plant utilization of nitrogen in dairy cattle manure. J Agric Sci 141, 79–91.

Van Soest, P.J., 1994. Nutritional ecology of the ruminant, 2nd ed. Cornell University Press, New York.

Tamminga, S., 2003. Pollution due to nutrient losses and its control in European animal production. Livest. Prod. Sci. 84, 101–111.

Taniwaki, M.H., Hocking, A.D., Pitt, J.I., Fleet, G.H., 2009. Growth and mycotoxin production by food spoilage fungi under high carbon dioxide and low oxygen atmospheres. Int. J. Food Microbiol. 132, 100–108.

Vance, E., Brookes, P., Jenkinson, D., 1987. An extraction method for measuring soil microbial biomass C. Soil Biol. Biochem. 19, 703–707.

Weete, J.D., Fuller, M.S., Huang, M.Q., Gandhi, S., 1989. Fatty acids and sterols of selected hyphochytriomycetes and chytridiomycetes. Exp. Mycol. 13, 183–195.

Weete, J.D., Weber, D.J., 1980. Lipid Biochemistry of Fungi and other Organisms. Plenum Press, New York.

Van Vliet, P.C.J., Reijs, J.W., Bloem, J., Dijkstra, J., de Goede, R.G.M., 2007. Effects of Cow Diet on the Microbial Community and Organic Matter and Nitrogen Content of Feces. J. Dairy Sci. 90, 5146–5158.

Wu, J., Joergensen, R.G., Pommerening, B., Chaussod, R., Brookes, P.C., 1990. Measurement of soil microbial biomass C by fumigation-extraction - an automated procedure. Soil Biol. Biochem. 22, 1167–1169.

Zelles, L., Hund, K., Stepper, K., 1987. Methoden zur relativen Quantifizierung der pilzlichen Biomasse im Boden. J Plant Nutr Soil Sci 150, 249–252.

6. Effect of cattle faeces with different microbial biomass content on soil properties, gaseous emissions and plant growth

Daphne Isabel Jost [ab*], Rainer Georg Joergensen [b], Albert Sundrum [a]

[a] Department of Animal Nutrition and Animal Health, University of Kassel, Nordbahnhofstr. 1a, 37213 Witzenhausen, Germany

[b] Department of Soil Biology and Plant Nutrition, University of Kassel, Nordbahnhofstr. 1a, 37213 Witzenhausen, Germany

Abstract

A study was carried out to investigate the effects of different diets for heifers, low and high yielding cows on the microbial composition of their faeces and subsequently the impacts of these faeces on CO_2 and N_2O emissions, N mineralisation and plant N uptake.

A diet low in N and high in ADF offered to heifers resulted in faeces dominated by fungi. These faeces were characterised by a low content in microbial biomass C and N and a high ergosterol concentration in comparison to the faeces of high yielding cows. Added to soil, faeces of heifers led to lower emission and stronger N immobilisation during a 14-day incubation in comparison to the faeces of high yielding cows. Total N_2O emission was significantly correlated with faecal microbial biomass N. Rye grass yield and N uptake were lowest in the soil supplemented with faeces from heifers in a 62-day pot experiment. Plant N uptake was influenced by the faecal microbial biomass C/N ratio and the fungal C to bacterial C ratio.

In conclusion, the faecal microbial biomass was affected to a high degree by the feeding regime and faecal microbial characteristics revealed higher impacts on plant N uptake than soil microbial properties.

Keywords: Cattle faeces; Microbial biomass C and N; Ergosterol, Amino sugar; Gaseous emission; N mineralisation

---

[*] Corresponding author. Tel.: + 49 5542 98 1523; e-mail: daphne.jost@arcor.de

6. Effect of cattle faeces with different microbial biomass content on soil properties, gaseous emissions and plant growth

6.1. Introduction

Solid manure from farm animals differs considerably in total N concentration and N availability to soil microorganisms and plants, due to differences in diet digestibility, diet conversion by different animal species, age of the animal, and water intake (Chadwick et al., 2000). In faeces, more than 90% of total N is present in organic forms (Jost et al., 2011), bound mainly in amino acids, but also in heterocyclic components (Bosshard et al., 2011) or amino sugars (Rezaeian et al., 2004a, b, 2006; Jost et al., 2011). Mineralisation of these organic forms is a prerequisite for plant uptake. Solid manure low in total N content and high in plant cell wall components has shown to result in lower net N mineralisation rates in laboratory incubation (Cusick et al., 2006; Morvan and Nicolardot, 2009; Peters and Jensen, 2011), greenhouse pot (Ikpe et al., 2003; Wu and Powell, 2007), and field experiments (Sørensen et al., 2003; Reijs et al., 2007; Powell and Grabber, 2009). The feeding regime also has significant impacts on the faeces-derived emissions of NH_3 (Merino et al., 2008; van der Stelt et al., 2008) and N_2O (Flessa et al., 2002a; Arriaga et al., 2010), causing serious N losses and atmospheric pollution.

Furthermore, differences in the chemical composition of the organic components affect the microbial community of faeces (van Vliet et al., 2007). Size and community structure of faecal microorganisms may control the decomposition of the faeces and the release of inorganic N after entering the soil (Chadwick et al., 2000). It is well-documented in various experiments with other organic material that the substrate colonising microbial community has an important influence on further decomposition processes (Flessa et al., 2002b) and directly adds significant amounts of microorganisms to the autochthonous soil microbial biomass (Rasul et al., 2008). Cattle faeces contain a highly dynamic community of bacteria, archaea and fungi which have not yet been qualified or quantified (Frostegård et al., 1997; Gattinger et al., 2007). Increased knowledge of the interactions between faecal composition and its behaviour in soil can lead to various strategies for livestock management and optimal utilisation, including method and time of application (Handayanta et al., 1997; Delve et al., 2001). As cattle diet and feed intake vary considerably between feeding regimes, influencing intestinal microbial parameters, direct measurements of the microbial community in faeces are an important source of information.

The objectives of this study were to assess the effects of different feeding regimes for heifers, low and high yielding cows on faecal composition in the first place and subsequently the impacts of the faeces on N_2O emissions, N mineralisation and plant N uptake. Faeces quality has previously been characterized microbially by determining microbial biomass C and N using the chloroform fumigation extraction (CFE) method, fungal ergosterol and the cell-wall components muramic acid and glucosamine (Jost et al., 2011). The CFE method is suited to differentiate accurately between living and dead microbial tissue (Brookes et al., 1985, Vance et al., 1987). Fungal ergosterol is an important component of fungal cell membranes (Weete and Weber, 1980). It has been repeatedly used as an index for fungal biomass in soil (Bååth and Anderson, 2003), other solid substrates (Newell, 1992), and in batch cultures of yeasts (Park et al., 1990). Fungal glucosamine and bacterial muramic acid were measured as independent control values for both CFE microbial biomass and ergosterol data (Appuhn and Joergensen, 2006; Indorf et al., 2011; Jost et al, 2011).

6.2. Material and methods

6.2.1. Soil

Soil samples were collected in March 2011 at 0-15 cm depth from the site "Saurasen", located in the north of Hesse, Germany. The site is at 280 m above sea level, with an average annual precipitation of 625 mm and temperature of 6.5 °C. Developed from eroded loess overlying clayey sandstone, the soil is classified according to the WRB classification system as Stagnic Luvisol (Quintern et al., 2006). The particle size distribution was 6 % sand, 72% silt and 22% clay. Soil organic C and total N content were 8.2 mg g^{-1} and 0.89 mg g^{-1}, respectively. The soil had a pH ($CaCl_2$) of 6.4, and a water holding capacity of 50%. After sampling, the soil was homogenized, sieved (<2 mm) and stored moist in polyethylene bags at room temperature for 18 days before the experiments started.

6.2.2. Faeces sampling and quality determination

Faeces samples were taken once at the same time from 18 cattle, six heifers, six high-level and six low-level yielding dairy cows (*Bos primigenius taurus*, var. German Holstein) from a cattle breeding farm in Lower Saxony. Heifers were fed *ad libitum* with a silage mix of grass and straw. Low and high yielding dairy cows (aprox. < 25 and 25-40 kg milk cow^{-1} d^{-1}, respectively) received the same silage within a total mixed ration (TMR), whereas the supplementation of concentrate differed with 5 and 9 kg d^{-1}, respectively. Crude protein content was highest in the diet of high yielding cows, followed by the low yielding cow fodder (data not shown). The feeding ration for the heifers contained more NDF than the other diets (491 versus 478 and 445 g kg^{-1} for low and high yielding cows, respectively). Faeces samples were taken rectally, immediately homogenised, frozen in liquid nitrogen and stored at -18 °C. This preservation technique has been previously tested, comparing the results of faecal analysis obtained this way with those of fresh samples, which were immediately analysed after sampling, showing no significant differences in the results between fresh and stored samples (data not shown). A subsample was dried for 72 h at 65 °C and finely ground for determination of faeces dry weight and chemical analyses. Total C and N, ammonium, and crude ash together with the organic components neutral detergent fibre (NDF), acid detergent fibre (ADF), acid detergent lignin (ADL), undigested dietary N (UDN), and undigested dietary C (UDC) were determined by near-infrared spectroscopy (FOSS 6500, Rellingen, Germany), as described by Althaus and Sundrum (University of Kassel, Witzenhausen, Germany, personal communication), after appropriate calibration and validation. Cellulose was calculated as the difference between ADF and ADL, hemicellulose as the difference between NDF and ADF. Easily decomposable carbon (C_E) and nitrogen (N_E) fractions in the feces were defined and calculated by subtraction of UDC and UDN from total faecal C and N, respectively. For further detail see Sundrum et al. (2011). All data are shown as the mean of six cows for each treatment.

6.2.3. Incubation experiment

The experiment comprised three soil treatments and one control: (1) faeces from heifers, (2) faeces from low yielding dairy cows, (3) faeces from high yielding dairy cows, and (4) control without faeces. Each treatment was replicated six times; faeces treatments comprised six individual faecal samples. The experiment was carried out in 1 l preserving jars. Each jar was filled with 50 g soil (on an oven-dry basis) at a bulk density of 1.0 g cm^{-3}. In the treatments with faeces addition, freshly thawed faeces were thoroughly mixed with the soil before filling into the jars. The application rate was 20 mg freshly thawed faeces g^{-1} soil dry weight (DW). This amount was equivalent to 0.4% of the soil DW and comprised an addition of 73 to 86 mg C per jar (18 – 21%) and 2.7 to 5.3 mg N per jar (6 – 12%).

The jars were incubated for 14 days at 22 °C and kept in the dark. At day 0, soil samples were removed from each jar to determine inorganic N. Soil respiration was measured as CO_2 and N_2O emissions. When starting the incubation, 10 ml 0.5 M NaOH were placed at the jar bottom and the lid was closed immediately. The NaOH solution was changed at day 3, 7 and 14. Evolved CO_2 was determined by back-titration to pH 8.3 of the excess NaOH with 0.5 M HCl after addition of 0.5 M $BaCl_2$ solution. N_2O emissions were measured at day 0, 1, 2, 3, 7, and 14. A 10 ml gas sample was taken out of the jar through a 3-layer silicone septum (Hamilton Company, Nevada, USA) with a plastic syringe. The samples were analysed immediately using a gas chromatograph GC-14B (Shimadzu Corporation, Kyoto, Japan). After 14 days, inorganic N and microbial biomass C and N were determined.

6.2.4. Pot experiment

The experiment had three soil treatments and one control: (1) faeces from heifers, (2) faeces from low-level dairy cows, (3) faeces from high-level dairy cows, and (4) control without faeces. Each treatment was replicated six times; faeces treatments comprised six individual faecal samples. The experiment was carried out in plastic pots (2.7 l, 13 × 13 cm). Each pot was filled with 2.4 kg soil (on an oven-dry basis) at a bulk density of 1.0 g cm^{-3}. In the three soil treatments, faeces were thoroughly mixed with the soil before filling into the pots. The application rate was 20 g freshly thawed faeces kg^{-1} soil. This amount was equivalent to 0.4% of the soil DW and comprised an addition of 3.5 to 4.1 g C per pot (1.4-1.7 g C kg^{-1} soil) and 130 to 252 mg N per pot (54-105 mg N kg^{-1} soil).

Italian ryegrass (*Lolium multiflorum*, breed Ligrande, from Deutsche Saatveredelung AG) was sown in a density of 250 seeds per pot (1.5 seeds cm^{-2}). The pots were arranged in a randomised complete block design and placed in a climate chamber with a 16-8 h light-dark cycle and peak temperatures of 20 °C (day) and 12 °C (night). Air humidity changed between 40% (day) and 90% (night). Soil moisture was kept at 50 % of the water-holding capacity by weighing and adding the water lost regularly three times a week. After germination, the first cut was conducted 5 cm aboveground at day 30 with a mean plant height of 15 cm. Total aboveground plants were harvested 62 days after germination. Samples were dried for 72 h at 40 °C for plant dry weight and finely ground to determine total C and N by combustion in a CNS Analyser (Elementar Vario EL, Elementar Analysensysteme GmbH Hanau, Germany). Mineral N was determined in soil sub-samples after harvesting.

6.2.5. Microbial biomass C and N

Microbial biomass C and N were estimated by the chloroform fumigation extraction method (Brookes et al., 1985, Vance et al., 1987). All soil samples were fumigated, extracted and measured for total C and N as described below. Two subsamples of 10 g fresh soil were taken for the analysis. One subsample was fumigated at 25 °C with ethanol-free CHCl$_3$, which was removed after 24 h. Fumigated and non-fumigated portions were extracted with 40 ml 0.05 M K$_2$SO$_4$ for 30 min by horizontal shaking at 200 rev min^{-1}. Soil extracts were filtered (folded filter paper, hw3, Sartorius Stedim Biotech, Göttingen, Germany). Organic C in the extracts was measured as CO$_2$ by infrared absorption after combustion at 850 °C using a Dimatoc 100 automatic analyser (Dimatec, Essen, Germany). Microbial biomass C was calculated as follows: microbial biomass C = E_C/k_{EC}, where E_C = (organic C extracted from fumigated faeces) − (organic C extracted from non-fumigated faeces) and k_{EC} = 0.45 (Wu et al., 1990; Joergensen, 1996). Total N in the extracts was measured by chemoluminescence detection. Microbial biomass N was calculated as follows: microbial biomass C = E_N/k_{EN}, where E_N = (organic N extracted from fumigated faeces) − (organic N extracted from non-fumigated faeces) and k_{EN} = 0.54 (Brookes et al., 1985; Joergensen and Mueller, 1996).

Faeces samples were fumigated and measured for C and N as described above, while the extraction was carried out according to the methodology described by Jost et al. (2011). Two freshly thawed subsamples equivalent to 0.5 g oven-dry faeces were taken for the analysis. Fumigated and non-fumigated portions were extracted with 100 ml 0.05 M $CuSO_4$ for 30 min by horizontal shaking at 200 rev min^{-1}. Following centrifugation (2000 g for 10 min), faeces extracts were filtered (hw3, Sartorius Stedim Biotech, Göttingen, Germany). C and N standards for calibration of the Dimatoc 100 analyser were prepared in 0.05 M $CuSO_4$ solution.

6.2.6. Ergosterol analysis

The fungal cell-membrane ergosterol was extracted and measured in soil according to Djajakirana et al. (1996). Moist soil samples of 0.5 g were extracted with 100 ml ethanol for 30 min by oscillating shaking at 250 rev min^{-1}. After filtering, the soil extract was evaporated in a vacuum rotary evaporator at 40 °C. The non-polar fraction was dissolved in 5 ml methanol and stored at 4 °C until measurement. For determination of ergosterol in faeces, the extraction method of Zelles et al. (1987) was used. Freshly thawed faeces equivalent to 0.5 g DW were placed into 30 ml test tubes and treated with 10 ml methanol, 2.5 ml ethanol and 1 g KOH. The sample was saponified for 90 min at 70 °C under reflux. After cooling, ergosterol was extracted in two steps with 15 + 10 ml petroleum ether. From the supernatant, 15 ml were evaporated in a vacuum rotary evaporator at 40 °C. The non-polar fraction was dissolved in 5 ml methanol and stored at 4 °C until measurement (Jost et al., 2011). Ergosterol was determined by reversed-phase HPLC with 100% methanol as the mobile phase and detected at a wavelength of 282 nm.

6.2.7. Amino sugar analysis

The amino sugars muramic acid, glucosamine and galactosamine were determined in faeces samples according to Indorf et al. (2011). Moist samples of 2 g fresh faeces were weighed into 20 ml test tubes, mixed with 10 ml 6 M HCl, and heated for 2 h at 105 °C. After HCl removal from the filtered hydrolysates in a vacuum rotary evaporator at 40 °C and centrifugation, the samples were transferred to vials and stored at -18 °C until the HPLC measurements. Chromatographic separations were performed on a Phenomenex (Aschaffenburg, Germany) Hyperclone C_{18} column (125 mm length × 4 mm diameter), protected by a Phenomenex C_{18} security guard cartridge (4 mm length × 2 mm diameter) at 35 °C. The HPLC system consisted of a Dionex (Germering, Germany) P 580 gradient pump, a Dionex Ultimate WPS – 3000TSL analytical autosampler with in-line split-loop injection and thermostat and a Dionex RF 2000 fluorescence detector set at 445 nm emission and 330 nm excitation wavelengths with medium sensitivity. OPA reagent (Merck Darmstadt), sample and preparation vials were stored in the autosampler at 15 °C. For the automated pre-column derivatisation, 50 µl OPA and 30 µl sample were mixed in the preparation vial and after 120 sec reaction time 15 µl of the indole derivates were injected. The mobile phase consisted of two eluents and was delivered at a flow rate of 1.5 ml min^{-1}. Eluent A was a 97.8/0.7/1.5 (v/v/v) mixture of an aqueous phase, methanol and tetrahydrofuran (THF). The aqueous phase contained 52 mmol sodium citrate and 4 mmol sodium acetate, adjusted to pH 5.3 with HCl. Then methanol and THF were added. Eluent B consisted of 50% water and 50% methanol (v/v).

Fungal C (mg g^{-1} dry weight) was calculated by subtracting bacterial glucosamine from total glucosamine as an index for fungal residues, assuming that muramic acid and glucosamine occur at a 1 to 2 molar ratio in bacterial cells (Engelking et al., 2007): mg fungal C g^{-1} dry weight = (mmol glucosamine − mmol muramic acid) × 179.2 g mol^{-1} × 9, where 179.2 is the molecular weight of glucosamine and 9 the conversion value of fungal glucosamine to fungal C (Appuhn and Joergensen, 2006). Bacterial C (µg g^{-1} dry weight) was calculated as an index for bacterial residues by multiplying the concentration of muramic acid in µg g^{-1} dry weight by 45 (Appuhn and Joergensen, 2006).

6.2.8. Inorganic N

Non-fumigated K$_2$SO$_4$ soil extracts were analysed for NH$_4^+$–N and NO$_3$–N by colorimetric analysis with a Continuous Flow Analyser (Evolution2, Alliance Instruments, Friedrichsdorf) at 540 nm. Net mineralised N from soil organic N in each treatment was calculated as the sum of the inorganic N forms at the end of the incubation time minus the initial inorganic N in soil.

6.2.9. Statistical analysis

The results presented in the tables are arithmetic means and expressed on an oven-dry basis (about 72 h at 60 °C). The significance of difference was tested by one-way analysis of variance. Statistical analyses were performed using JMP 7.0 (SAS Inst. Inc.).

6.3. Results

6.3.1. Differences in faeces characteristics

Differences in the feeding regime between the three treatments resulted in clear differences in the composition of their faeces (Table 1). The concentration of total N, the C/N ratio, N_E and hemicellulose content differed significantly between the faeces treatments. In contrast, the concentration of UDN, ADF and cellulose showed no significant differences between treatments. C_E was higher in the faeces of heifers than in the other groups. In faeces of high yielding cows, the faecal C/N ratio was lower with a higher dietary N concentration compared to the other treatments. Concerning faecal chemical properties, negative correlation coefficients were found between total N and ADF ($r = -0.72$, $P < 0.001$) and between N_E and ADL ($r = -0.80$, $P < 0.0001$).

Table 1. Elemental composition and organic components in different cattle feeding regimes and in cattle faeces

	N	NH_4^+	C/N	N_E	UDN	UDC	NDF	ADF	Cellulose	Hemic.	Crude ash
	(mg g^{-1} DW)						(mg g^{-1} DW)				
Feeding (TMR)											
Heifer	21						491	342	267	149	185
Low yielding	21						478	246	240	232	106
High yielding	25						445	214	234	231	92
Faeces											
Heifer	16 a	1.7 ab	28 a	0.93 a	6.3 a	25 a	472 a	383 a	281 a	89 a	196 a
Low yielding	22 b	1.5 a	20 b	1.52 b	7.1 a	29 a	512 ab	379 a	288 a	133 b	158 b
High yielding	28 c	2.1 b	16 c	2.17 c	6.8 a	30 b	522 b	353 a	281 a	169 c	150 b
CV (±%)	12	19	10	17	9.3	3.4	6.0	5.8	5.4	16	6.5

N_E = easily decomposable nitrogen, UDN = undigested dietary nitrogen, UDC = undigested dietary carbon, NDF = neutral detergent fibre, ADF = acid detergent fibre, Hemic. = hemicellulose; DW = dry weight, TMR = total mixed ration, CV = pooled coefficient of variation between feeding regimes (cow replicates n = 6); different letters indicate a significant difference (Tukey/Kramer, $P < 0.05$).

Microbial biomass C, microbial biomass N, and bacterial muramic acid were highest in the faeces of high yielding cows in comparison to the faeces of heifers ($P < 0.05$) (Table 2). Microbial indices revealed no significant differences between the faeces of low and high yielding cows, except for ergosterol. This fungal biomarker was significantly lower in the faeces of low yielding cows than in the other faeces. The concentration of glucosamine, galactosamine and microbial C did not differ between faeces of the three treatments. Faecal microbial biomass C/N ratio and the ratio of fungal C to bacterial C showed a strong positive correlation with $r = 0.79$ ($P < 0.001$). Both ratios significantly declined with increasing N content (Table 3). The correlation coefficients were $r = -0.76$ and $r = -0.74$, respectively ($P < 0.001$). The ratio ergosterol to microbial biomass C was lowest in the faeces of heifers, which in contrast revealed the highest ratio of fungal glucosamine to ergosterol. Faecal UDC correlated negatively with the microbial C to N ratio ($r = -0.74$, $P = 0.0004$) and with the fungal C to bacterial C ratio ($r = -0.88$, $P < 0.0001$). Similarly, an increase in N_E went along with a lower portion of fungi in relation to bacteria, while higher C_E values correlated positively with a higher fungal portion, but also with muramic acid ($r = 0.73$, $P = 0.0007$).

Table 2. Microbial biomass indices, amino sugar indices and microbial C in cattle faeces from different feeding regimes

Faeces type	Microbial biomass C	N	Ergosterol	MurN	ManN	GlcN	GalN	Microbial C
	(mg g^{-1} DW)		(µg g^{-1} DW)	(mg g^{-1} DW)				
Heifer	10 a	1.3 a	4.4 a	0.22 a	0.07 a	1.5 a	0.79 a	21 a
Low yielding	14 ab	2.0 ab	2.2 b	0.41 b	0.15 b	2.0 a	1.15 a	31 a
High yielding	18 b	3.1 b	3.8 a	0.45 b	0.09 ab	2.0 a	0.83 a	32 a
CV (±%)	22	25	23	23	49	26	29	14

DW = dry weight; MurN = muramic acid; ManN = mannosamine; GlcN = glucosamine; GalN = galactosamine; CV = pooled coefficient of variation between feeding regimes (cow replicates n = 6); different letters indicate a significant difference (Tukey/Kramer, $P < 0.05$)

Table 3. Microbial indices in cattle faeces from different feeding regimes

Faeces type	Microbial biomass C/N	Ergosterol/ microbial biomass C(‰)	Fungal C/ bacterial C	Fungal glucosamine/ ergosterol
Heifer	7.9 a	0.45 a	1.10 a	296 a
Low yielding	6.8 b	0.15 b	0.66 b	672 b
High yielding	5.6 c	0.27 ab	0.60 b	385 ab
CV (±%)	12	11	11	12

CV = pooled coefficient of variation between feeding regimes (cow replicates n = 6); different letters indicate a significant difference (Tukey/Kramer, $P < 0.05$).

6. Effect of cattle faeces with different microbial biomass content on soil properties, gaseous emissions and plant growth

6.3.2. Effects of faeces types on soil microorganisms and grass growth

The addition of the cattle faeces to the soil significantly increased the contents of soil microbial biomass N and fungal ergosterol (Table 4). However, increase in soil microbial biomass C was only significant in the case of faeces from the low yielding cows. Faeces addition increased the amounts of CO_2 evolved. CO_2 emission was highest when faeces of the high yielding cows were added in comparison with the other two faeces types ($P < 0.05$). The release of CO_2-C from faeces was negatively correlated with the faecal microbial biomass C/N ratio ($r = -0.69$; $P = 0.002$) and positively with soil ergosterol ($r = 0.78$, $P < 0.002$). Addition of faeces to soil generally led to immobilisation of inorganic N, i.e. negative net N mineralisation. N immobilisation in soil was significantly highest when faeces of heifers were supplemented, compared with the faeces of the other two treatments. Faeces addition generally increased N_2O emission rates. A clear increase of N_2O emission occurred especially in the treatment with faeces of high yielding cows (Fig. 1). Total N_2O emission was significantly correlated with faecal microbial biomass N ($r = 0.58$; $P < 0.05$, $n = 24$).

Table 4. Microbial indices, CO_2 emissions and mineral nitrogen in different soil treatments after 14 days of incubation

Soil treatment	Microbial biomass		Ergosterol	Respiration CO_2-C	N_{min}
	C	N	($\mu g\ g^{-1}$ soil)		
Heifer	227 ab	62 a	0.64 a	246 a	-8.6 a
Low yielding	261 a	71 a	0.71 a	262 a	-5.5 b
High yielding	234 ab	64 a	0.67 a	330 b	-4.3 b
Control	187 b	36 b	0.38 b	52 c	3.2 c
CV (±%)	14	24	18	10	26

CV = pooled coefficient of variation between feeding regimes (n = 6); different letters indicate a significant difference (Tukey/Kramer, $P < 0.05$).

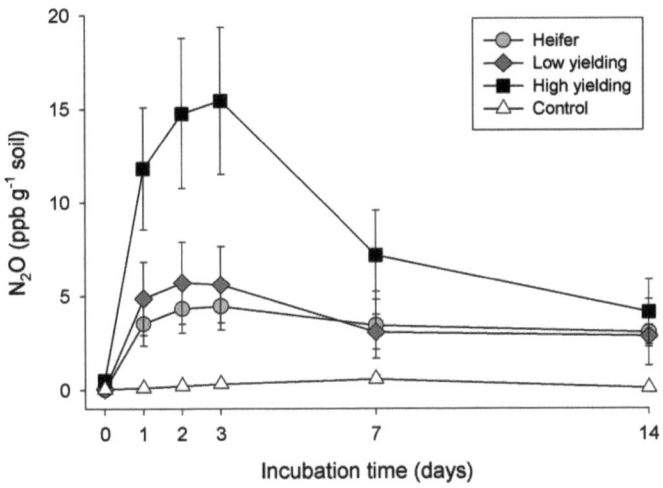

Fig. 1. N_2O emissions from different soil treatments during 14 days of incubation at 22°C; bars indicate ± one standard error; n = 6.

In the pot experiment, inorganic N was immobilised in all treatments, lowest in the control and highest in the treatment with faeces of heifers. Consequently, highest rye grass yield and N uptake was found in the control treatment without addition of faeces, lowest in the treatment with faeces of heifers (Fig. 2). However, the differences in faecal composition revealed no significant differences in N immobilisation. Plant N uptake showed the closest linear relationship with the faecal N_E and total N content ($r = 0.84$ and 0.81, $P < 0.001$), followed by a negative relationship with the faecal microbial biomass C/N ratio ($r = -0.76$, $P < 0.001$). Also the faecal fungal C to bacterial C ratio showed a significant negative relationship with the plant N uptake.

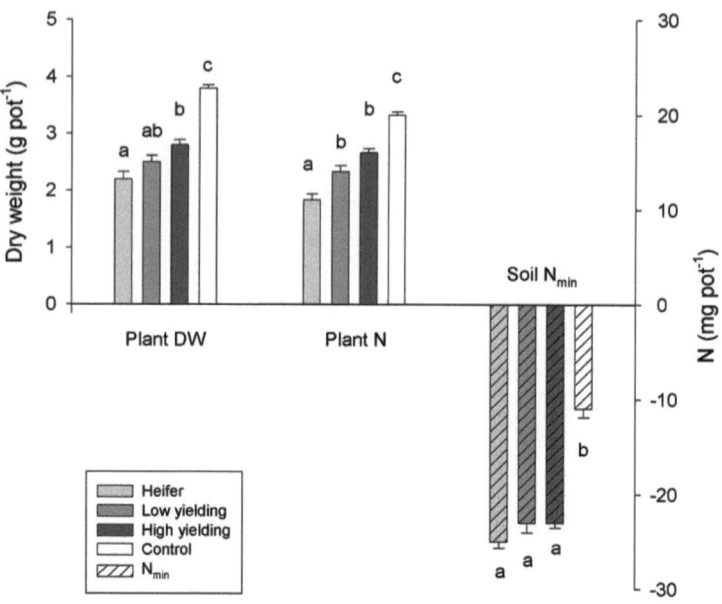

Fig. 2. Harvest of Italian ryegrass after growing for 62 days on different soil treatments. Plant dry weight per pot, plant nitrogen content and mineral nitrogen immobilised from the soil per pot; bars indicate ± one standard error; different letters indicate a significant difference (Tukey/Kramer, $P < 0.05$); n = 6.

6.4. Discussion

6.4.1. Microbial indices

The concentrations of all microbial indices were in the middle of the range described by Jost et al. (2011) in differently fed dairy cows and heifers. The range of the data for microbial biomass C, N, and ergosterol obtained by these two studies was larger than that for the amino sugars muramic acid, glucosamine, galactosamine, and mannosamine. These differences between the studies are most likely due to differences in diet composition. An N balanced or protein rich diet, low in crude fibre, NDF and ADF generally supports faecal microbial and especially faecal bacterial biomass (van Vliet et al., 2007). This might also be the reason for the different contribution of microbial biomass C obtained by fumigation extraction to total microbial C, calculated on the basis of bacterial muramic acid and fungal glucosamine. According to Amelung (2001, 2008), as cell-wall components, amino sugars have the tendency to be accumulated in microbial residues, suggesting that roughly 50% of the faecal microbial C in the present study belong to the living fraction and the other 50% are within the remains of dead fungi and bacteria. Jost et al. (2011) reported that the living fraction contributed roughly 30%. It cannot be completely excluded that the low percentage was caused by the different sample treatments, shock-freezing in liquid N_2 in the present study and slow freezing at -18 °C by Jost et al. (2011).

In the highly dynamic situation of C and N supply in the gut, rapid microbial growth is probably similarly accompanied by concomitant microbial death. However, the differences between the microbial indices estimates by fumigation extraction and amino sugar analysis should not be stressed too much, because it is not known (1) whether the glucosamine and muramic acid concentrations are identical in living and dead microbial tissue or (2) whether the several conversion values for microbial biomass C by the fumigation extraction method and for microbial C by amino sugar analysis may lead to an underestimation of microbial biomass C (Jost et al., 2011). The comparison is even more complex as archaea contribute a significant percentage to the microbial biomass of cattle faeces. Gattinger et al. (2007) estimated that the archaeal phospholipid etherlipids added approximately 16% to the total phospholipid chain content in dairy cow manure. Archaea most likely add to the fraction of $CHCl_3$ labile C. However, they do not contain muramic acid, but galactosamine, glucosamine, and a variety of other rare amino sugars such as N-acetyl-L-talosaminuronic acid (Kandler and König, 1998).

6.4.2. Bacterial and fungal contribution to microbial tissue

In the current study, the mean faecal concentration of ergosterol was 3.5 µg g^{-1} DW. Klamer and Bååth (2004) obtained a factor of 190 when calculating the fungal biomass C from the ergosterol concentration in 11 compost fungi species. Taking this factor into account would result in a mean fungal biomass C content of 0.67 mg C g^{-1} DM in the present faeces samples, leading to the suggestion that ergosterol containing fungi contributed a portion of 5% to total microbial biomass C, which is in the range obtained by Jost et al. (2011). However, this value is considerably below the value obtained by amino sugar analysis. Of the anaerobic fungal populations, yeasts contain high concentrations of ergosterol (Aguilera et al., 2006; Ahmad et al., 2010). In contrast, no ergosterol but high concentrations of cholesterol were measured in chytridiomycetes (Weete et al., 1989; Kagami et al., 2007). Information is still lacking on the ergosterol concentration of anaerobic fungal species found in the cattle rumen such as *Anaeromyces, Orpinomyces, Caecomyces,* or *Piromyces* (Griffith et al., 2009).

6. Effect of cattle faeces with different microbial biomass content on soil properties, gaseous emissions and plant growth

On the other hand, amino sugar analysis, comprising all fungi, revealed a higher proportion of fungi in the current study. The fungal C to bacterial C ratio of 0.60 to 1.1 observed corresponds to a fungal contribution of between 37% and 52% to microbial C. These percentages are similar or slightly above the values obtained by Jost et al. (2011) for heifers from the same farm. Chitin, the polymer of glucosamine, has been previously used as an indicator for fungal biomass in rumen fluid of cattle (Sekhavati et al., 2009) and in sheep faeces (Rezaeian et al., 2004a, b). The chitin concentration in sheep faeces was 10.2 mg g^{-1} DW (Rezaeian et al., 2004a). One reason for the markedly higher chitin concentration in comparison to the glucosamine concentration in the present study might be the differences in rumen and intestine physiology between sheep and cattle. However, Rezaeian et al. (2004a, 2006) used the colorimetric assay of Chen and Johnson (1983), which cannot distinguish between fungal glucosamine, bacterial glucosamine, and galactosamine. This may have led to an overestimation of fungal tissue.

6.4.3. Effects of feeding regime and faeces composition

Feeding regimes affected not only the chemical but also the microbial composition of cattle faeces. A diet low in N and high in ADF diet which was offered to heifers resulted in faeces dominated by fungi (calculated from fungal glucosamine). The faeces were low in microbial biomass C and N in combination with a wide microbial biomass C/N ratio and a high ergosterol concentration, especially in comparison with the diet of high lactating cows. This effect is common as a higher proportion of roughage in the feed promote fungal growth (Rezaeian et al., 2006) because fungi prefer C-rich cell wall material and need less N concentration in the substrate.

Gaseous emissions from faeces added to soil were significantly influenced by the composition and characteristics of faeces. Release of CO_2-C and N_2O was highest in the soil amendment with faeces of high yielding cows, providing a higher amount of easy available C and N substrates than the other faeces. This is known to increase gaseous emissions (Flessa et al., 2002a; Arriaga et al., 2010).

Added to soil, the faeces of heifers, characterised by low N and high fibre content, led to a stronger N-immobilisation than the other treatments. An increased microbial biomass C/N ratio in combination with an increased fungal C to bacterial C ratio indicates that the shift in the microbial community structure towards fungi had a depressive effect on the N storage of faecal microorganisms, i.e. less N is incorporated into microbial components. In soil, an increasing microbial biomass C/N ratio together with an increasing fungal biomass has not often been found (Heinze et al., 2010; Joergensen et al., 2010).

The N immobilisation found in this study has been repeatedly observed after direct application of cattle faeces to soil (Griffin et al., 2005; Cusick et al., 2006; Peters and Jensen, 2011). This was to be expected, as its C/N ratio of 28 was somewhat above the threshold of 25, indicating restricted N availability to soil microorganisms (Powlson et al., 2001). However, the strong N immobilisation in the faeces of high yielding cows is surprising, considering the fact that the C/N ratio was far below 20, where an N release to the soil solution usually occurrs (Janssen, 1996; Seneviratne, 2000; Peters and Jensen, 2011). In contrast to the N-free faecal organic C fraction, the faecal organic N fractions seem hardly to be available to soil microorganisms, caused by the strong microbial decomposition during gut passage. Nevertheless, solid-state cross polarisation ^{15}N nuclear magnetic resonance spectroscopy and Curie-point pyrolysis–gas chromatography/mass spectrometry also failed to give clear further information on the chemical reasons for the poor availability of faecal organic N to soil microorganisms and consequently to plants (Bosshard et al., 2011). The authors assumed that some N compounds present in plants were not digested in the gut, and that also some excreted recalcitrant N compounds were de novo synthesized by gut microorganisms. An N fixation capacity like in the present cattle faeces has been observed for highly decomposed sugarcane filter pressmud, which led to N immobilisation even at an initial C/N ratio of 12 (Rasul et al., 2008). In the long-term, N from immobilising faeces was also released by the microbial decomposition processes into the soil solution (Morvan and Nicolardot, 2009; Peters and Jensen, 2011) and taken up by plants (Chadwick et al., 2000).

N uptake by rye grass increased with faecal N concentration and decreased with higher faecal crude fibre content. Plant N uptake decreased also with a higher fungal C to bacterial C ratio and a higher microbial biomass C/N ratio. A negative relationship between diet composition, faecal N concentration and N mineralisation has been repeatedly observed for sheep faeces (Kyvsgaard et al., 2000), but also cattle faeces (Sørensen et al., 2003; Wu and Powell, 2007; Morvan and Nicolardot, 2009). However, none of these investigations considered the microbial quality of faeces. A rare and interesting exception was the study of van Vliet et al. (2007), which was focussed on faecal bacteria. Nevertheless, closer relationships have been observed between plant N uptake and faecal microbial properties than of plant N uptake and soil microbial properties. The microbial biomass C/N ratio and the fungal C to bacterial C ratio in cattle faeces had a strong negative impact on plant N uptake. Highest CO_2 evolution and lowest N immobilisation suggest a stronger turnover of the microbial biomass in the faeces treatment with high N and N_E concentration and low ADF concentration. The increased microbial turnover might be the reason why the increase in microbial biomass did not consistently differ between the three faeces types. The relationships between faecal and soil processes might be masked by the autochthonous microbial community. The use of ^{13}C and ^{15}N labelled cattle faeces would make it possible to test these assumptions using an identical approach to that presented in this study (Sørensen and Jensen, 1998; Jensen et al., 1999; Bosshard et al, 2011; Wachendorf and Joergensen, 2011). Hence, it would be possible to investigate wether microbially incorporated C or N derive from faeces or soil.

6.4.4. Conclusions

This study showed that a feeding regime low in protein and high in fibre content resulted in less microbial, but higher fungal biomass. Gaseous emissions from faeces added to soil were significantly influenced by the composition and characteristics of faeces, but also by the faecal microbial indices. Composition of faeces showed a clear impact on N-mineralisation in soil and consequently the N supply to plants, faecal microbial properties revealing closer relationships to plant N uptake than soil microbial properties. This is a factor which is not taken into account frequently. The considerable impact of faeces derived microorganisms on emission potential and plant growth in this study requires further investigations. Besides, the addition of faeces from heifers with a high C/N ratio and a microbial community dominated by fungi may contribute to the soil N supply in the longer run, which should also be considered. Continuative studies will help to identify the best management practices to reduce gaseous emissions and will also allow farmers to develop good practices for efficient nutrient use.

Acknowledgments

The skilful technical assistance of Gabriele Dormann and Christiane Jatsch is highly appreciated. This project was supported by a grant of the Research Training Group 1397 "Regulation of soil organic matter and nutrient turnover in organic agriculture" of the German Research Foundation (DFG).

6.5. References

Aguilera, F., Peinado, R.A., Millán, C., Ortega, J.M., Mauricio, J.C., 2006. Relationship between ethanol tolerance, H^+-ATPase activity and the lipid composition of the plasma membrane in different wine yeast strains. Int. J. Food Microbiol. 110, 34–42.

Ahmad, A., Khan, A., Manzoor, N., Khan, L.A., 2010. Evolution of ergosterol biosynthesis inhibitors as fungicidal against Candida. Microb. Pathog. 48, 35–41.

Amelung, W., 2001. Methods using amino sugars as markers for microbial residues in soil. In: Lal, J.M., Follett, R.F., Stewart, B.A. (Eds.), Assessment Methods for Soil Carbon. Lewis Publishers, Boca Raton, pp. 233-272.

Amelung, W., Brodowski, S., Sandhage-Hofmann, A., Bol, R., 2008. Combining biomarker with stable isotope analyses for assessing the transformation and turnover of soil organic matter. Adv. Agron. 100, 155-250.

Appuhn, A., Joergensen, R.G., 2006. Microbial colonisation of roots as a function of plant species. Soil Biol. Biochem. 38, 1040–1051.

Arriaga, H., Salcedo, G., Calsamiglia, S., Merino, P., 2010. Effect of diet manipulation in dairy cow N balance and nitrogen oxides emissions from grasslands in northern Spain. Agric. Ecosyst. Environ. 135, 132-139.

Bååth, E., Anderson, T.H., 2003. Comparison of soil fungal/bacterial ratios in a pH gradient using physiological and PLFA-based techniques. Soil Biol. Biochem. 35, 955-963.

Bosshard, C., Oberson, A., Leinweber, P., Jandl, G., Knicker, H., Wettstein, H.-R., Kreuzer, M., Frossard, E., 2011. Characterization of fecal nitrogen forms produced by a sheep fed with ^{15}N labeled ryegrass. Nutr. Cycl. Agroecosyst. 90, 355-368.

Brookes, P.C., Landman, A., Pruden, G., Jenkinson, D.S., 1985. Chloroform fumigation and the release of soil nitrogen: a rapid direct extraction method for measuring microbial biomass nitrogen in soil. Soil Biol. Biochem. 17, 837-842.

Chadwick, D.R., John, F., Pain, B.F., Chambers, B., Williams, J., 2000. Plant uptake of nitrogen from the organic nitrogen fraction of animal manures: a laboratory experiment. J. Agr. Sci. 134, 159–168.

Chen, G.C., Johnson, B.R., 1983. Improved colorimetric determination of cell wall chitin in wood decay fungi. Appl. Environ. Microbiol. 46, 13–16.

Cusick, P.R., Powell, J.M., Kelling, K.A., Hensler, R.F., Muñoz, G.R., 2006. Dairy manure N mineralization estimates from incubations and litterbags. Biol. Fertil. Soils 43, 145–152.

Delve, R.J., Cadisch, G., Tanner, J.C., Thorpe, W., Thorne, P.J., Giller, K.E., 2001. Implications of livestock feeding management on soil fertility in the smallholder farming systems of sub-Saharan Africa. Agric. Ecosyst. Environ. 84, 227–243.

Djajakirana, G., Joergensen, R., Meyer, B., 1996. Ergosterol and microbial biomass relationship in soil. Biol. Fertil. Soils 22, 299–304.

Engelking, B., Flessa, H., Joergensen, R.G., 2007. Shifts in amino sugar and ergosterol contents after addition of sucrose and cellulose to soil. Soil Biol. Biochem. 39, 2111-2118.

Flessa, H., Ruser, R., Dörsch, P., Kamp, T., Jimenez, M.A., Munch, J.C., Beese, F., 2002a. Integrated evaluation of greenhouse gas emissions (CO_2, CH_4, N_2O) from two farming systems in southern Germany. Agric. Ecosyst. Environ. 91, 175-189.

Flessa, H., Potthoff, M., Loftfield, N., 2002b. Laboratory estimates of CO_2 and N_2O emissions following surface application of grass mulch: importance of indigenous microflora of mulch. Soil Biol. Biochem. 34, 875–879.

Frostegård, A., Petersen, S.O., Bååth, E., Nielsen, T.H., 1997. Dynamics of a microbial community associated with manure hot spots as revealed by phospholipid fatty acid analyses. Appl. Environ. Microbiol. 63, 2224–2231.

Gattinger, A., Höfle, M.G., Schloter, M., Embacher, A., Böhme, F., Munch, J.C., Labrenz, M., 2007. Traditional cattle manure application determines abundance, diversity and activity of methanogenic archaea in arable European soil. Environ. Microbiol. 9, 612–624.

Griffin, T.S., He, Z., Honeycutt, C.W., 2005. Manure composition affects net transformation of nitrogen from dairy manures. Plant Soil 273, 29-38.

Griffith, G.W., Ozkose, E., Theodorou, M.K., Davies, D.R., 2009. Diversity of anaerobic fungal populations in cattle revealed by selective enrichment culture using different carbon sources. Fungal Ecol. 2, 87-97.

Handayanta, E., Cadisch, G., Giller, K.E., 1997. Regulating N mineralization from plant residues by manipulation of quality. In: Cadisch, G., Giller, K.E. (Eds.), Driven by Nature - Plant Litter Quality and Decomposition. CAB International, Wallingford, UK, pp. 175–186.

Heinze, S., Raupp, J., Joergensen, R.G., 2010. Effects of fertilizer and spatial heterogeneity in soil pH on microbial biomass indices in a long-term field trial of organic agriculture. Plant Soil 328, 203–215.

Ikpe, F.N., Ndegwe, N.A., Gbaraneh, L.D., Torunana, J.M.A., Williams, T.O., Larbi, A., 2003. Effects of sheep browse diet on fecal matter decomposition and N and P cycling in the humid lowlands of West Africa. Soil Sci. 168, 646-659.

Indorf, C., Dyckmans, J., Khan, K.S., Joergensen, R.G., 2011. Optimisation of amino sugar quantification by HPLC in soil and plant hydrolysates. Biol. Fertil. Soils 47, 387–396.

Janssen, B.H., 1996. Nitrogen mineralization in relation to C:N ratio and decomposability of organic materials. Plant and Soil 181, 39–45.

Jensen, B., Sørensen, P., Thomsen, I.K., Jensen, E.S., Christensen, B.T., 1999. Availability of nitrogen in ^{15}N-labeled ruminant manure components to successively grown crops. Soil Sci. Soc. Am. J. 63, 416-423.

Joergensen, R.G., 1996. Quantification of the microbial biomass by determining ninhydrin-reactive N. Soil Biol. Biochem. 28, 301–306.

Joergensen, R.G., Mueller, T., 1996. The fumigation-extraction method to estimate soil microbial biomass: calibration of the k_{EN} value. Soil Biol. Biochem. 28, 33-37.

Joergensen, R.G., Mäder, P., Fließbach, A., 2010. Long-term effects of organic farming on fungal and bacterial residues in relation to microbial energy metabolism. Biol. Fertil. Soils 46, 303-307.

Jost, D.I., Indorf, C., Joergensen, R.G., Sundrum, A., 2011. Determination of microbial biomass and fungal and bacterial distribution in cattle faeces. Soil Biology and Biochemistry 43, 1237–1244.

Kagami, M., von Elert, E., Ibelings, B.W., De Bruin, A., Van Donk, E., 2007. The parasitic chytrid, *Zygorhizidium*, facilitates the growth of the cladoceran zooplankter, *Daphnia*, in cultures of the inedible alga, *Asterionella*. Proc. Royal Soc. B: Biol. Sci. 274, 1561-1566.

Kandler, O., König, H., 1998. Cell wall polymers in Archaea (Archaebacteria). Cell. Mol. Life Sci. 54, 305–308.

Klamer, M., Bååth, E., 2004. Estimation of conversion factors for fungal biomass determination in compost using ergosterol and PLFA 18:2ω6,9. Soil Biol. Biochem. 36, 57–65.

Kyvsgaard, P., Sørensen, P., Møller, E., Magid, J., 2000. Nitrogen mineralization from sheep faeces can be predicted from the apparent digestibility of the feed. Nutr. Cycl. Agroecosyst. 57, 207–214.

Merino, P., Arriaga, H., Salcedo, G., Pinto, M., Calsamiglia, S. 2008. Dietary modification in dairy cattle: field measurements to assess the effect on ammonia emissions in the Basque country. Agric. Ecosyst. Environ. 123, 88–94.

Morvan, T., Nicolardot, B., 2009. Role of organic fractions on C decomposition and N mineralization of animal wastes in soil. Biol. Fertil. Soils 45, 477-486.

Newell, S.Y., 1992. Estimating fungal biomass and productivity in decomposing litter. In: Carroll, G.C., Wicklow, D.T. (Eds.), The Fungal Community. Its Organization and Role in the Ecosystem, 2nd Edition. Marcel Dekker, New York. pp. 521-561.

Park, J.-W., Lee, W.-S., Bang, W.-G., 1990. The production of ergosterol by *Saccharomyces sake*. KBA No. 6. J. Korean Agric. Chem. Soc. 33, 87-92.

Peters, K., Jensen, L.S., 2011. Biochemical characteristics of solid fractions from animal slurry separation and their effects on C and N mineralisation in soil. Biol. Fertil. Soils 47, 447-455.

Powell, J.M., Grabber, J.H., 2009. Dietary forage impacts on dairy slurry nitrogen availability to corn. Agron. J. 101, 747-753.

Powlson, D.S., Hirsch, P.R., Brookes, P.C., 2001. The role of soil microorganisms in soil organic matter conservation in the tropics. Nutrient Cycling in Agroecosystems 61, 41–51.

Quintern, M., Lein, M., Joergensen, R.G., 2006. Changes in soil biological quality indices after long-term addition of shredded shrubs and biogenic waste compost. J. Plant Nutr. Soil Sci. 169, 488-493

Rasul, G., Khan, K.S., Müller, T., Joergensen, R.G., 2008. Soil-microbial response to sugarcane filter cake and biogenic waste compost. J. Plant Nutr. Soil Sci. 181, 355-360.

Reijs, J.W., Sonneveld, M.P.W., Sørensen, P., Schils, R.L.M., Groot, J.C.J., Lantinga, E.A., 2007. Effects of different diets on utilization of nitrogen from cattle slurry applied to grassland on a sandy soil in The Netherlands. Agric., Ecosyst. Environ. 118, 65-79.

Rezaeian, M., Beakes, G.W., Parker, D.S., 2004a. Distribution and estimation of anaerobic zoosporic fungi along the digestive tracts of sheep. Mycol. Res. 108, 1227–1233.

Rezaeian, M., Beakes, G.W., Parker, D.S., 2004b. Methods for the isolation, culture and assessment of the status of anaerobic rumen chytrids in both in vitro and in vivo systems. Mycol. Res. 108, 1215–1226.

Rezaeian, M., Beakes, G.W., Chaudhry, A.S., 2006. Effect of feeding chopped and pelleted lucerne on rumen fungal mass, fermentation profiles and in sacco degradation of barley straw in sheep. Anim. Feed Sci. Technol. 128, 292–306.

Sekhavati, M.H., Mesgaran, M.D., Nassiri, M.R., Mohammadabadi, T., Rezaii, F., Maleki, A.F., 2009. Development and use of quantitative competitive PCR assays for relative quantifying rumen anaerobic fungal populations in both in vitro and in vivo systems. Mycol. Res. 113, 1146–1153.

Seneviratne, G., 2000. Litter quality and nitrogen release in tropical agriculture: a synthesis. Biol. Fertil. Soils 31, 60–64.

Sørensen, P., Jensen, E.S., 1998. The use of ^{15}N labelling to study the turnover and utilization of ruminant manure N. Biol. Fertil. Soils 28, 56–63.

Sørensen, P., Weisbjerg, M.R., Lund, P., 2003. Dietary effects on the composition and plant utilization of nitrogen in dairy cattle manure. J. Agric. Sci. 141, 79–91.

Sundrum, A., E. Schlecht, R. G. Joergensen (2011): Fractions of nitrogen and carbon in feces of lactating dairy cows and their variability on dairy farms. J. Dairy Sci. (accepted).

van der Stelt, B., van Vliet, P.C.J., Reijs, J.W., Temminghoff, E.J.M., Van Riemsdijk, W.H. 2008. Effects of dietary protein and energy levels on cow manure excretion and ammonia volatilisation. J. Dairy Sci. 91, 4811-4821.

van Vliet, P.C.J., Reijs, J.W., Bloem, J., Dijkstra, J., de Goede, R.G.M., 2007. Effects of cow diet on the microbial community and organic matter and nitrogen vontent of feces. J. Dairy Sci. 90, 5146–5158.

Vance, E., Brookes, P., Jenkinson, D., 1987. An extraction method for measuring soil microbial biomass C. Soil Biol. Biochem. 19, 703–707.

Wachendorf, C., Joergensen, R.G., 2011. Mid-term tracing of ^{15}N derived from urine and dung in soil microbial biomass. Biol. Fertil. Soils 47, 147-155.

Weete, J.D., Weber, D.J., 1980. Lipid Biochemistry of Fungi and other Organisms. Plenum Press, New York.

Weete, J.D., Fuller, M.S., Huang, M.Q., Gandhi, S., 1989. Fatty acids and sterols of selected hyphochytriomycetes and chytridiomycetes. Exp. Mycol. 13, 183–195.

Wu, J., Joergensen, R.G., Pommerening, B., Chaussod, R., Brookes, P.C., 1990. Measurement of soil microbial biomass C by fumigation-extraction-an automated procedure. Soil Biol. Biochem. 22, 1167–1169.

Wu, Z., Powell, J.M., 2007. Dairy manure type, application rate, and frequency impact plants and soils. Soil Sci. Soc. Am. J. 71, 1306-1313.

Zelles, L., Hund, K., Stepper, K., 1987. Methoden zur relativen Quantifizierung der pilzlichen Biomasse im Boden. J. Plant Nutr. Soil Sci. 150, 249–252.

7. Zusammenfassung

Um die Nährstoffeffizienz im Boden zu erhöhen und klimarelevante Gasemissionen zu minimieren, sind quantitative Informationen zu den C- und N-Fraktionen im Kot sowie deren Mineralisierungspotential nötig. Da über die Hälfte des fäkalen N mikrobiell gebunden ist, sind Methoden zur Bestimmung der mikrobiellen Kotqualität hilfreich.

Ziele der ersten Publikation waren die Anwendung der CFE-Methode zur Bestimmung der mikrobiellen Biomasse in Rinderkot, Ergosterolbestimmung als Marker für die pilzliche Biomasse und Aminozuckernachweis zur Analyse der mikrobiellen Gemeinschaftsstruktur (pilzliches Glucosamin und bakterielle Muramin-säure). Mit Hilfe der CFE-Methode sollten lebende Mikroorganismen im Kot, inklusive Bakterien, Pilze und Archaeen, erfasst werden. Verschiedene Extraktionsmittel wurden für diese Methode getestet, um stabile Extrakte und reproduzierbare mikrobielle Biomasse-C- und -N-Gehalte zu erhalten. Der Einsatz von 0.05 M $CuSO_4$ als Extraktionsmittel löste vorherige Probleme mit der Extraktion $CHCl_3$-labiler N-Komponenten und sorgte für stabile Kotextrakte. Die Methoden wurden in einem Kotinkubationsexperiment bei 25 °C verglichen. Mikrobielle Parameter zeigten dynamische Charakteristika und mögliche Verschiebungen innerhalb der mikrobiellen Gemeinschaft. Im Kot von Färsen betrug das mittlere C/N-Verhältnis 5,6 und der mittlere C_{mik}/C_{org}-Quotient 2,2%, das Verhältnis von Ergosterol zum mikrobiellen Biomasse-C war 1,1‰. Ergosterol und Aminozuckeranalyse ergaben einen signifikanten Pilzanteil von über 40% des mikrobiellen Gesamt-C. Für die Analyse mikrobieller Parameter in Rinderkot erwiesen sich alle getesteten Methoden als geeignet. Diese wurden für die folgenden Fütterungsversuche weiter unabhängig voneinander angewendet und verglichen.

Die zweite Publikation verglich eine N-defizitäre (ND) und eine ausgeglichene N-Bilanz (NB) bei der Fütterung von Milchkühen unter Berücksichtigung der Kot-inhaltsstoffe, der mikrobiellen Parameter und der Verdaulichkeit. Unterschiede zwischen Individuen und Probennahmetagen wurden ebenfalls miteinbezogen. Mittlerer mikrobieller Biomasse-C- und -N-Gehalt war 37 bzw. 4,9 mg g^{-1} TM. Der Pilzanteil lag diesmal bei 25% des mikrobiellen Gesamt-C. Die Fütterung zeigte signifikante Effekte auf die Kotzusammensetzung. Das fäkale C/N-Verhältnis der NB-Fütterung war signifikant niedriger als bei ND. Gleiches gilt für das C/N-Verhältnis der mikrobiellen Biomasse mit jeweils 9.1 und 7.0 für ND und NB. Auch die Verdaulichkeit wurde durch die Fütterung beeinflusst. Unverdauter Futterstickstoff, Faserstoffe (NDF) und Hemicellulose waren in der ND-Behandlung signifikant erhöht. Einige Parameter zeigten nur einen Einfluss der Probennahmetage, mit den angewendeten Methoden gelang jedoch der eindeutige

7. Zusammenfassung

Nachweis der Fütterungseffekte auf mikrobielle Parameter im Rinderkot, wobei sich die Fütterung in nur einer Variable unterschied.

Weitere Fütterungseinflüsse auf die Kotqualität wurden schließlich auch für Rinder unterschiedlicher Leistungsstufen erforscht. Hier waren die Unterschiede in der Fütterung wesentlich größer als in den vorhergehenden Experimenten. Der Kot von Färsen, Niederleistungs- und Hochleistungskühen sowie dessen Einfluss auf N_2O-Emissionen, N-Mineralisierung und pflanzliche N-Aufnahme wurden untersucht. Die Färsenfütterung mit geringem N- und hohem ADF-Anteil führte zu pilzdominiertem Kot. Besonders im Vergleich zum Kot der Hochleistungskühe war der Gehalt an mikrobiellem Biomasse-C niedrig, in Verbindung mit einem breiten mikrobiellen C/N-Verhältnis und hohem Ergosterolgehalt. Eingemischt in Boden zeigte Färsenkot die niedrigste CO_2-Produktion und höchste N-Immobilisierung sowie niedrigste N_2O-Emissionen während einer 14-tägigen Inkubation bei 22 °C.

In einem 62-Tage-Gefäßversuch mit Welschem Weidelgras waren der Trocken-masseertrag und die pflanzliche Stickstoffaufnahme in den Färsenbehandlungen am niedrigsten. Die Stickstoffaufnahme durch die Pflanzen korrelierte positiv mit der Stickstoffkonzentration im Kot und negativ mit dem Rohfasergehalt, aber auch negativ mit dem C/N-Verhältnis der mikrobiellen Biomasse und dem Verhältnis von pilzlichem zu bakteriellem C. Mikrobielle Parameter im Kot zeigten einen größeren Zusammen-hang mit der pflanzlichen Stickstoffaufnahme als die mikrobiellen Bodenparameter.

8. Summary

As an important component of organic fertilizers, animal faeces require methods for determining diet effects on their microbial quality to improve nutrient use efficiency in soil and to decrease gaseous greenhouse emissions to the environment. Since more than half of the faecal nitrogen originates from microbial N, a knowledge of the microbial biomass content in livestock faeces is useful. The objectives of the first study were to apply the chloroform fumigation extraction (CFE) method for determining microbial biomass in cattle faeces, to determine the fungal cell-membrane component ergosterol, and to measure the cell-wall components fungal glucosamine and bacterial muramic acid as indices for the microbial community structure. The CFE method should detect living organisms, including bacteria, fungi and achaea. A variety of extractant solutions were tested for the CFE method to obtain stable extracts and reproducible microbial biomass C and N values. Application of 0.05 M $CuSO_4$ as extractant solved previous problems with the extraction of $CHCl_3$ labile N components and provided stable faeces extracts. The methods were compared in a 28-day cattle faeces incubation study at 25 °C. Here, the microbial biomass indices showed dynamic characteristics and possible shifts in the microbial community. In faeces of five heifers, the mean microbial biomass C/N ratio was 5.6, the mean microbial biomass to organic C ratio was 2.2%, and the mean ergosterol to microbial biomass C ratio was 1.1‰. Ergosterol and amino sugar analysis revealed a significant contribution of fungi, with a percentage of more than 40% of the total microbial C. All methods applied are expected to be suitable tools for analysing the microbial quality of cattle faeces. They were adopted independently and further compared for the following feeding experiments.

In the second publication, an N deficient (ND) and an N balanced diet (NB) were compared with respect to the impacts on faecal composition, microbial indices and digestibility. Differences between individuals and sampling days were also considered. The mean values of microbial biomass C and N concentrations averaged around 37 and 4.9 mg g-1 DM, respectively. Ergosterol, together with fungal glucosamine and bacterial muramic acid, revealed a contribution of fungi in dairy cattle faeces of 25% fungal C as a percentage of total microbial C. Changes in ruminal N supply showed significant effects on faecal composition. Faecal C/N ratio of the NB treatment was significantly lower than in ND. The same applies for the mean microbial biomass C/N ratio, which differed significantly, at 9.1 and 7.0 for ND and NB, respectively. Digestibility was also significantly affected by the feeding regime. Undigested dietary N, fibre fractions (NDF) and hemicellulose were significantly higher in ND compared with the NB treatment. There were

8. Summary

parameters showing only an effect of the sampling days. However, with the methods applied, clear feeding effects on microbial parameters in cattle faeces were confirmed.

Further feeding effects on fecal quality were investigated for cattle of different performance levels. A study was carried out to investigate first the effects of different diets for heifers, low and high lactation cows on their faeces and subsequently the effects of these different cattle faeces types on N_2O emissions, N mineralisation and plant N uptake. A low N and high ADF diet given to heifers resulted in a faeces dominated by fungi, which was low in microbial biomass C in combination with a wide microbial biomass C/N ratio and a high ergosterol concentration, especially in comparison with the diet of high lactating cows. Added to soil, heifer faeces led to the lowest microbial CO_2 production and strongest N immobilisation but also to lowest N_2O emissions during a 14-day incubation at 22 °C. Also, the rye grass yield and plant N uptake was lowest in the soil amend with heifer faeces in a 62-day pot experiment. N uptake by rye grass correlated positively with the faecal N concentration and negatively with its crude fibre concentration, but also negatively with the microbial biomass C/N ratio and the fungal C to bacterial C ratio. Faecal microbial properties revealed closer relationships to plant N uptake than soil microbial properties.

9. Schlussfolgerung und Ausblick

Die CFE-Methode eignet sich zur Bestimmung des mikrobiellen Biomasse-C und -N in Rinderkot. Hierbei bietet sie den Vorteil, nur lebende Mikroorganismen nachzuweisen, wobei neben Bakterien und Pilzen auch die Archeenbiomasse erfasst wird. In Kombination mit pilzlichem Ergosterol deutet die CFE-Methode auf Verschiebungen innerhalb der mikrobiellen Gemeinschaftsstruktur im Rinderkot während des Alterungsprozesses. Ergosterol und Aminozuckeranalyse ergaben einen Pilzanteil von über 40% des mikrobiellen Gesamtkohlenstoffs in Färsenkot. Im Methodenvergleich zeigten die CFE-Methode und Aminozuckeranalyse ähnliche Ergebnisse, obwohl pilzliches Glucosamin und bakterielle Muraminsäure auf einen hohen Anteil toter Mikroorganismen im Kot hinwiesen. Der mikrobielle Biomasseanteil konnte jedoch durch Kotkonservierung in flüssigem Stickstoff erhöht werden.

Mit guter Wiederholbarkeit der Methoden auch bei verschiedenen Fütterungen oder Leistungsstufen eignen sich die angewendeten Methoden auch im Bereich der Tierernährung und Tiergesundheit, um das mikrobielle Verhalten im Kot oder anderem Darminhalt zu verfolgen. Besonders Ergosterol und Aminozuckeranalyse gaben Hinweise auf das Bakterien-Pilzverhältnis und damit auf deren möglicher Beteiligung an der Verdauung. Ein Stickstoffmangel im Futter bewirkte ein höheres mikrobielles Biomasse-C/N-Verhältnis und eine geringere Verdauung von Faseranteilen. Die Bestimmung mikrobieller Parameter lieferte nützliche Informationen über Fütterungs-effekte auf die Rinderkotbeschaffenheit und deren Düngungspotential. Die Variabilität der Kotzusammensetzung zwischen Individuen und Probennahmetagen muss hierbei statistisch berücksichtigt werden.

Nährstoffverluste durch Gasemissionen aus dem Boden nach Kotzugabe wurden stark durch das Futter beeinflusst. Kot von Hochleistungsmilchkühen mit einem hohen Kraftfutteranteil im Futter zeigte die höchsten CO_2-C und N_2O-Emissionen. Rinderkot-zugabe kann im Boden zu N-Immobilisierung führen, die bei einem hohen Faser- und niedrigen Stickstoffanteil im Futter der Färsen am stärksten ausgeprägt war. Die fäkale N-Aufnahme durch Pflanzen wurde besonders durch den N-Gehalt im Futter beeinflusst. Im Gefäßversuch zeigte sich eine vermehrte Aufnahme bei hoher N-Zufuhr, während ein höherer fäkaler Faseranteil die N-Aufnahme der Pflanzen erniedrigte. Das Verhältnis von Bakterien und Pilzen sowie das des mikrobiellen Biomasse-C und -N zeigte ebenfalls eine negative Korrelation mit Pflanzen-N. Hoher Faser- und niedriger N-Gehalt im Futter führten zu einem Anstieg des Pilzanteils im Kot. Der Zusammen-hang zwischen pflanzlicher N-Aufnahme und mikrobiellen Parametern im Kot war größer als mit denen im Boden. Die angewandten Methoden eignen sich deshalb für die Analyse von Fütterungseffekten auf

9. Schlussfolgerung und Ausblick

mikrobielle Parameter in Rinderkot sowie deren Verhalten im Boden. Das Forschungsprojekt sollte mit ^{15}N- und ^{13}C-markiertem Kot fortgesetzt zu werden, um die genauen Nährstoffquellen für Mikroorganismen und Pflanzen herauszufinden (Sørensen and Jensen, 1998; Jensen et al., 1999; Bosshard et al, 2011; Wachendorf and Joergensen, 2011).

Die luminometrische Bestimmung des ATP-Gehaltes bleibt viel versprechend. Durch Vorinkubation der Extrakte nach Enzymzugabe wurde inzwischen eine Messwertstabilität erreicht, die nur noch geringe Abweichungen zeigt. Die Methode sollte mit dem letzten Versuchsansatz nach Redmile-Gordon (2011) und mit in Stickstoff konservierten Proben fortgesetzt werden. Für weitere Versuche mit Rinderkot ist außerdem eine Konservierung in Flüssigstickstoff ratsam.

10. Literatur

Amelung, W., 2001. Methods using amino sugars as markers for microbial residues in soil. In: Lal, J.M., Follett, R.F., Stewart, B.A. (Eds.), Assessment Methods for Soil Carbon. Lewis Publishers, Boca Raton, pp. 233-272.

Amelung, W., Brodowski, S., Sandhage-Hofmann, A., Bol, R., 2008. Combining biomarker with stable isotope analyses for assessing the transformation and turnover of soil organic matter. Advance in Agronomy 100, 155-250.

Anderson, J.P.E., Domsch, K.H., 1978. A physiological method for quantitative measurement of microbial biomass in soils. Soil Biol. Biochem. 10, 519-525

Appuhn, A., Joergensen, R.G., Raubuch, M., Scheller, E., Wilke, B., 2004. The automated determination of glucosamine, galactosamine, muramic acid and mannosamine in soil and root hydrolysates by HPLC. J Plant Nutr Soil Sci 167, 17-21.

Bosshard, C., Oberson, A., Leinweber, P., Jandl, G., Knicker, H., Wettstein, H.-R., Kreuzer, M., Frossard, E., 2011. Characterization of fecal nitrogen forms produced by a sheep fed with ^{15}N labeled ryegrass. Nutr. Cycl. Agroecosyst. 90, 355-368.

Brookes, P.C., Landman, A., Pruden, G., Jenkinson, D.S., 1985. Chloroform fumigation and the release of soil nitrogen: A rapid direct extraction method for measuring microbial biomass nitrogen in soil. Soil Biol. Biochem. 17, 837-842

Brookes, P.C., Newcombe, A.D., Jenkinson, D.S., 1987. Adenylate energy charge measurements in soil. Soil Biol. Biochem. 19, 211-217.

Van Bruchem, J., Verstegen, M. W., Tamminga, S. 2000. From nutrient fluxes in animals in nutrient dynamics and health in animal production systems. EAAP-Publ. 97, 28-48.

Diez-Gonzales, F., Callaway, T. R., Kizoulis M. G., Russel J., B., 1998. Grain feeding and the dissemination of acid-resistant *Escherichia coli* from cattle. Science 281, 1666-1669.

Djajakirana, G., Joergensen, R., Meyer, B., 1996. Ergosterol and microbial biomass relationship in soil. Biol. Fertil. Soils 22, 299–304.

Dyckmans, J., Raubuch, M., 1997. A modification of a method to determine adenosine nucleotides in forest organic layers and mineral soils by ion-paired reversed-phase high-performance liquid chromatography. J. Microbiol. Methods 30, 13-20.

Engelking, B., Flessa, H., Joergensen, R.G., 2007. Shifts in amino sugar and ergosterol contents after addition of sucrose and cellulose to soil. Soil Biol.Biochem. 39, 2111-2118.

10. Literatur

Firestone, M. K., Davidson, E., A., 1989. Microbiological basis of NO and N_2O production and consumption in soil. In: Andreae M.,O., Schimel D., S. (Eds.) Exchange of Trace Gases Between Terrestrial Ecosystems and the Atmosphere. John Wiley & Sons Ltd., Chichester, pp. 7-21.

Flessa, H., Dörsch, P., Beese, F., 1995. Seasonal variation of N_2O and CH_4 fluxes in differently managed arable soils in Southern Germany. J. Geophys. Res. 100, 23115-23124.

Gruber, L., 1990. Proteinversorgung von Milchkühen. Bericht über die 17. Tierzucht-tagung „Aktuelle Fragen der Milchvieh- und Schafhaltung", BAL Gumpenstein, pp. 29-36

Harmon, B.G., Brown, C.A., Tkalcic, S., Mueller, P.O.E., Parks, A., Jain, A.V., Zhao, T., Doyle, M.P., 1999. Faecal shedding and rumen growth of *E. coli* O157:H7 in fasted calves. J. Food Prot. 62, 574-579

Höper, H., Kleefisch, B., 2001. Untersuchung bodenbiologischer Parameter im Rahmen der Boden-Dauerbeobachtung in Niedersachsen – Bodenbiologische Referenzwerte und Zeitreihen. Arbeitshefte Boden 2001/4, NLfB, Hannover

IPCC, 1997. Intergovernmental Panel on Climate Change guidelines for national green-house gas inventories, chapter 4, Agriculture: nitrous oxide from agricultural soils and manurte management, OECD, Paris, France.

IPCC, 2001. Good practice guidance and uncertainty management in national green-house gas inventories. Intergovernmental Panel on Climate Change, www.ipcc.ch.

Jenkinson, D.S., 1966. Studies on the decomposition of plant material in soil. II. Partial sterilization of soil and the soil biomass. J. Soil Sci. 17, 280-302.

Jenkinson, D.S., 1988. The determination of microbial biomass carbon and nitrogen in soil. In: Wilson, J.R., (Ed.), Advances in Nitrogen Cycling in Agricultural Ecosystems. CABI, Wallingford, pp. 368-386.

Jenkinson, D.S., Oades, J.M.. 1979. A method for measuring adenosine triphosphate in soil. Soil Biol. Biochem. 11, 193-199.

Jensen, B.B., Jørgensen, H., 1994. Effect of dietary fiber on microbial activity and microbial gas production in various regions of the gastrointestinal tract of pigs. Appl. Envir. Microbiol., 60, 1897-1904.

Jensen, B., Sørensen, P., Thomsen, I.K., Jensen, E.S., Christensen, B.T., 1999. Availability of nitrogen in ^{15}N-labeled ruminant manure components to successively grown crops. Soil Sci. Soc. Am. J. 63, 416-423.

Jörgensen, R.G., 1995. Die quantitative Bestimmung der mikrobiellen Biomasse in Böden mit der Chloroform-Fumigations-Extraktionsmethode. Gött. Bodenkundl. Ber. 104, 1-229

Joergensen, R.G., 2000. Ergosterol and microbial biomass in the rhizosphere of grassland soils. Soil Biol. Biochem. 32, 647-652.

Kalk, W. D., Hülsbergen, K. J., 1996. Methodik zur Einbeziehung des indirekten Energieverbrauchs mit Investitionsgütern in Energiebilanzen von Landwirtschafts-betrieben. Kühn-Archiv 90, 41-56.

Lindecrona, R.H., Jensen, T.K., Jensen, B.B., Leser, T.D., Jiufeng, W., Møller, K., 2003. The influence of diet on the development of swine dysentery upon experimental infection. Animal Sci. 76, 81-87

Mäder, P., Fliessbach, A., Dubois, D., Gunst, L., Fried, P., Niggli, U., 2002. Soil fertility and biodiversity in organic farming. Science 296,1694-1697.

Oades, J.M. and Jenkinson, D.S., 1979. Adenosine triphosphate content of the soil microbial biomass. Soil Biol. Biochem. 11 , 201–204.

Redmile-Gordon, M., White, R.P., Brookes, P.C., 2011. Evaluation of substitutes for paraquat in soil microbial ATP determinations using the trichloroacetic acid based reagent of Jenkinson and Oades 1979. Soil Biol. Biochem. 43, 1098-1100.

Salzgeber, C., Lörcher, M., 1997. Produktökobilanz Brot unter verschiedenen Landbaubedingungen. In: Deutsche Bundesstiftung Umwelt (Ed) Umweltverträgliche Pflanzenproduktion – Indikatoren, Bilanzierungsansätze und ihre Einbindung in Öko-Bilanzen. Zeller-Verlag, Osnabrück, pp. 249-269.

Sehy U., 2004. N_2O-Freisetzung landwirtschaftlich genutzter Böden unter dem Einfluss von Bewirtschaftungs- und Standortfaktoren. Dissertation, TU-München-Weihen-stephan

Singh, J. S., Raghubanshi, A. S., Singh, R. S., Srivastava, S. C., 1989. Microbial biomass acts as a source of plant nutrients in dry tropical forest and savanna. Nature 338, 499-500.

Sørensen, P., Jensen, E.S., 1998. The use of ^{15}N labelling to study the turnover and utilization of ruminant manure N. Biol. Fertil. Soils 28, 56–63.

Vandermeer, J., 1995. The ecological basis of alternative agriculture. Ann. Re.v Ecol. Syst. 26, 201-224.

Vance, E., Brookes, P., Jenkinson, D., 1987. An extraction method for measuring soil microbial biomass C. Soil Biol. Biochem. 19, 703–707.

Verhoeven, F. P. M., van der Made, G. M., van Bruchem, J., 1998. Nitrogen in balance – results of 100 Friesian dairy farms show the importance of the uptake from the soil. Veeteelt 15, 496-498.

Wachendorf, C., Joergensen, R.G., 2011. Mid-term tracing of ^{15}N derived from urine and dung in soil microbial biomass. Biol. Fertil. Soils 47, 147-155.

Weete, J.D., Weber, D.J., 1980. Lipid Biochemistry of Fungi and other Organisms. Plenum Publishing, New York.

Wolstrup, J., Jensen, K., 2008. Adenosine triphosphate and deoxyribonucleic acid in the alimentary tract of cattle fed different nitrogen sources. Journal of Applied Microbiology 45, 49–56.

10. Literatur

Zelles, L., Hund, K., Stepper, K., 1987. Methoden zur relativen Quantifizierung der pilzlichen Biomasse im Boden. J. Plant Nutr. Soil Sci. 150, 249–252.

Zhang, X., Amelung, W., 1996. Gas chromatographic determination of muramic acid, glucosamine, mannosamine, and galactosamine in soils. Soil Biol. Biochem. 28, 1201-1206.

11. Danksagung

Als erstes möchte ich mich für die erstklassige Betreuung meiner Arbeit bedanken:

Herrn Prof. A. Sundrum danke ich für seine Kompetenz, sein Verständnis und seine fachmännische Beratung in Ernährungsfragen. Er gewährte mir spannende Einblicke in die Geheimnisse des ruminalen Verdauungstraktes.

Herrn Prof. R. G. Jörgensen danke ich für seinen ausgezeichneten fachkundigen Rat sowie für seine Geduld und Ruhe mit der er als Fels in der Brandung Doktoranden vor der Verzweiflung zu bewahren vermag.

Außerdem danke ich Herrn Prof. B. Ludwig und Frau Dr. habil. K. Michel, die mir nicht nur ein DFG-Stipendium, sondern auch ein faszinierendes Projekt anvertraut haben.

Ich danke der DFG für die Finanzierung und die Möglichkeit, unter optimalen Bedingungen zu forschen, zu entdecken und zu publizieren.

Der Abteilung Bodenbiologie danke ich für die besten Arbeitskollegen:

Mein besonderer Dank gilt Gabriele Dormann, unsere technisch geniale Laborfee und geheime Projektkoordinatorin, die uns in der Not Seelentrost spendete. Liebe Gabi, ich danke dir für deine Geduld, deine Nerven und deine Person, dir verleihe ich die Ehrendoktorwürde.

Ein großes Dankeschön an Susanne Beck, unsere Heldin, die in schimmernder Rüstung den Kampf mit allen widrigen Elementen, welche ein Fortkommen zu behindern drohen, leidenschaftlich aufnimmt. Dich, liebe Susi, erhebe ich in den Ritterstand.

Allen Mitdoktoranden: Stefanie Heinze, Nicole Heyn, Caroline Indorf, Ramia Jannoura, Stefan Lukas, Rajasekaran Murugan, Nils Rottmann, André Sradnick, Charlotte Tönnshoff, Stefanie Wentzel - danke für euer geteiltes Leid, Mitleid, die schönste Mittagszeit, für eure Freizeit und Fröhlichkeit, danke für die tolle Zeit.

Allen Azubis: Ann-Katrin Becker, Nicole Gaus, Sophie Trümper, Sabine Schröter, Sabine Werk, Matthias Wollrath - danke für eure Hilfe und euren Fleiß mit all den leckeren Proben und danke fürs Mitnehmen, fürs Feiern und für eure Ohren - es hat Spaß gemacht mit euch.

Außerdem: Margit Rode - danke für deinen Rat, deinen Enthusiasmus und deinen schwarzen Humor.

11. Danksagung

Meiner Bürokollegin, Ramia Jannoura, bin ich überaus dankbar für eine perfekte Symbiose zwischen absoluter Wohlfühl- und angenehmer Arbeitsatmosphäre. Liebe Ramia, ich würde mir auch heute mit niemand anderem lieber ein Büro teilen als mit dir.

Ein großes Dankeschön gebührt auch der Abteilung für Tierernährung:

Den Mitarbeitern: Christiane Jatsch, Charlotte Marien, Birgül Althaus, Nicole Gaus, Ulrich Sünder, Susanne Hoischen-Taubner, Simona Büchel, Uwe Richter - danke für euren fachlichen Rat, eure Hilfe im Labor, bei der Probennahme und für die Mitnahme.

Einen herzlichen Dank an die Mitarbeiter des Wentrothofes für die gute Zusammen-arbeit. Und danke an die Beteiligten des FLI Braunschweig für ihre Hilfe und ihren Rat. Natürlich danke ich auch allen Färsen und Milchkühen für ihre großzügige Spende.

Big thanks to Mr Brookes and his advice. He told me that my PhD will be the best time of my life and, despite all problems and disappointments with certain methods, he was right.

Schließlich möchte ich meiner Familie danken, die mir diese Arbeit ermöglicht und immer hinter mir gestanden hat:

Meiner Schwester Johanna - danke fürs Daumendrücken. Meinem Bruder Patrick - danke für den Kaffee, ohne den hätte ich's nicht geschafft.

Meinen geliebten Eltern - danke für eure Unterstützung, fürs Essen einpacken, für die Englischkorrektur in der Not, für euer Verständnis und eure Liebe.

Mein lieber Rudi, meine liebe Walli - danke für ein zweites Zuhause und für den Platz in eurem Herzen.

Meiner Fu, die mir beim Schreiben zu Hause mit absoluter Gelassenheit zur Seite stand und mir auch in schwierigen Zeiten Aufmunterung garantierte.

Meinem geliebten Bernd - danke für deine Geduld und den geistigen Beistand, die mich soweit gebracht haben. Als bärenstarker Kämpfer verdienst du die Lorbeeren. Ich danke dir von ganzem Herzen.

i want morebooks!

Buy your books fast and straightforward online - at one of world's fastest growing online book stores! Environmentally sound due to Print-on-Demand technologies.

Buy your books online at
www.get-morebooks.com

Kaufen Sie Ihre Bücher schnell und unkompliziert online – auf einer der am schnellsten wachsenden Buchhandelsplattformen weltweit! Dank Print-On-Demand umwelt- und ressourcenschonend produziert.

Bücher schneller online kaufen
www.morebooks.de

 VDM Verlagsservicegesellschaft mbH
Heinrich-Böcking-Str. 6-8   Telefon: +49 681 3720 174   info@vdm-vsg.de
D - 66121 Saarbrücken   Telefax: +49 681 3720 1749   www.vdm-vsg.de

Printed by Books on Demand GmbH, Norderstedt / Germany